FROM HERE TO INFINITY

Bruce D Jimerson
BS, MS, JD

A world without Time

From the dawn of History, humans have obsessed over the nature of *"time."* The earliest written records express confusion, anxiety and frustration as to its momentary and fleeting existence. Much of Greek philosophy was directed to making sense out of ones transience on the earth. The scientific approach to *"time"* shrugs aside the wisdom of traditional cultures by which *"durations"* are known intuitively. Before Galileo, *"time"* was a subjective thing, not a parameter to be measured with precision. Newton extracted *"time"* from the 17th century world, and gave it an abstract independent existence as a way of keeping track of motion; it did not serve any other function other than as an accounting of vehicle. Three centuries later, Einstein restored *"time"* to the world, going so far as to give temporal intervals a dimensional status on par with spatial increments. Minkowsky quickly elaborated upon Einstein's work by unifying space and time into a single entity, spacetime.

Einstein's *"time"* and Minkowsi's *"spacetime"* provided valuable conceptual tools that allowed physicists to explore a strange and counterintuitive world. To explain the unsuccessful attempts of experimenters to detect the speed of light relative to the earth, Einstein reasoned that space, time and relative motion would be interdependent in such a way that every observer would measure the speed of light as equal to '**c**, irrespective of the observers motion wrt the space-frame in which the light wave was propagating. Special Relativity appeared inconsistent with the earlier work of James Clerk Maxwell relating propagation speed to the electrical and magnetic properties of the void. The bizarre and shocking consequence of Einstein's theory, was that time does not pass at the same rate for everyone. Special Relativity, however, was predicted upon changes in both space and time; it could still fit within the constrains of Maxwell's Theory.

The question remained, however, as to whether it was a true account of nature. When evidence of time dilations were observed for high speed particles traveling in the earth's atmosphere, the scientific community jumped to the conclusion that Einstein's special theory had been verified. But these were preferred frame experiments -- referenced to a clock on the earth taken to be at rest in a non-moving inertial frame. To be compatible with Maxwell's Theory, both space and time had to be altered from the perspective of a clock in the other frame. So while hi-speed clocks were observed to run slow relative to earth clocks, there were no experiments to establish the reciprocal proposition that earth clocks ran slow when observed from the perspective of the high speed clock. Nor were there any experiments evidencing the reality of spatial contraction. Nonetheless, Special Relatively became standard doctrine, that is, until the Hafele-Keeting experiment and GPS snags provoked new questions that were better answered by treating "time dilations" as energy differences between relatively moving inertial frames.

What had prevailed as the 20th century idea of *"time"* as a flow that could be altered by energy, was to be later embraced within the idea that *"time"* began at the instant of a big bang. Having no prior existence, as an entity concurrent with the emergence of space, one is immediately compelled to ask the unanswerable question - what happened before time begin? To put the question, is to contradict the premise. Indeed, if there be no time, all is space, changing space. No past and no future, neither a beginning nor an end. In lieu of clocks, we must look to the growth of space to ratify the illusion of temporal reality.

Forward

To chronicle existence without creation, is to embrace the notion of space as impedance and gravitation as reactance, for that is the way the world is made. Vacuum divergence is to gravity what changing momentum is to inertial matter. Behind it all, a subliminal dynamic. To comprehend the cosmos, one must think inside the expanding box.

Contents

Forward 5

Spatial Inertia..7

Geometrical Dynamics of Inertial Bodies 8

Transformation of Hubble Mass to Infinite Plane.......13

Inertial Space as Area Density Continuum................ 17

The Emergence of Gravity 18

Gravity as Pressure 19

Second law Symmetry 20

Origin of Virtual inertia 22

Recapitulation 27

Spatial Inertia

Sir Isaac Newton's recognition that falling apples were governed by the same law as orbiting planets, was perhaps his most far sighted and controversial contribution to the scientific world. That inertial objects could reach across empty space to exert attractive forces upon one another, was considered by many "voodoo physics."

It would be more than two centuries before the world received a tenable alternative. Albert Einstein realized masses do not act directly upon one another. Rather, inertial matter according to Einstein, was deemed to modify both space and time. General Relativity replaced gravitational force with spacetime distortion. But the theory did not explain how mass bent space, nor did it predict the origin or intensity of the **G** field. Comes then Alexander Friedmann, a Russian mathematician heralding a new interpretation of G.R. The universe could be explained in terms of spatial dynamics. Although receiving little attention at the time, Friedmann's opus would become of great interest following Hubble's discovery of expansion. Yet the cause of gravity remained a puzzling enigma. It was Richard Feynman, who insightfully suggested ... perhaps gravity was an inertial reaction.

Historically, the path that begins with Newton's "Law of Attraction" between separated masses, is initially augmented by General Relativity, then later reformed as a dynamic by Friedmann, thence relegated to a common pseudo force by Feynman. Still, there was no comprehensive theory of gravity -- Feynman did not identify the primary agent of action (the spatial acceleration field required to create an inertial counter force). To return full circle back to Newton by way of his 2nd Law of Motion, the pseudo force must be orchestrated by a pressure creating acceleration field. And finally, Newton's second law must by symmetrical (i.e., the force created by the acceleration of mass wrt space must be equal to the force created by the acceleration of space wrt mass.

In *Standard quantum Theory*, the influence of separated bodies upon one another relies upon unseen "go-between particles" invented, by their advocates, as momentum transferring entities. The herein developed theory of the void as an inertial dynamic, relies not upon invisible particles, but rather a hitherto overlooked apotheosis of spatial expansion. Momentum is transferred by spatial expansion pressure. By this ambit, all things are spatially connected. To be shown, **g** fields are instantiated by four concurrences:

1) Expansion created isotropic spatial acceleration

2) The inertial opposition of mass opposing expansion per (1).

3) The action of the pseudo force field created by (2) upon the universe.

4) The convergence of the cosmic pseudo force field created (3) upon the mass.

Within what is known of the extent of space and the mass of matter, the cosmos is a superlative vacuum, exceeding that attainable in laboratories. Yet an accelerated or decelerated mass in free space feels the cosmic presence in some fashion as opposition. By what manner can space bring about gravity, inertial reaction and the anomalies of propagation?

Surprisingly, the existence of such an artifice has long been known, but commonly viewed as having little cosmological applicability – indeed, the miraculous properties of infinite laminae(s) were considered largely academic. That all changed in the latter years of the 20th century. Coming with the 1998 supernova studies, were the imperatives of exponential expansion, infinite space and the problem of finding sufficient energy to fund accelerated expansion.

While neither Hubble mass nor size is separately utile, taken together they define the area density of a plane representative of the universe in all matters pertaining to inertia and gravity. Newton's 2nd Law, in defining the dimensionality of inertial force in terms of *mass, time* and *space,* set the stage for formulating gravity as the unavoidable consequence thereof.

Geometrical Dynamics of Inertial Bodies

As a constellation of distributed inertial matter, one might surmise the Hubble sphere could only behave operationally as a rarefied volume. With respect to uniformly moving massive bodies far removed from gravitational influences, chance encounters with a few hydrogen atoms per meter will have nil affect upon velocity. But this is distinctly not the case for accelerating bodies. Any change in momentum, whether increasing or decreasing, is instantly opposed. By what manner can virtually empty space exert acceleration opposing forces?

A perfect vacuum, by definition, is empty. The average density ρ_U of the Hubble sphere in 3-D form as thinly distributed lumps of matter is approximately = 3×10^{-26} **kg/m³** (an unattainable vacuum by laboratory means). Yet this virtually empty volume, nearly devoid of mass and substance, inaugurates counter forces proportional to the mass of an accelerating object, bafflingly independent of the volume, area, shape or uniformity thereof?

A rigid Hubble universe would exhibit a bare mass inertia in the range **1.5 x10⁵³ kg**. Such a structure is immediately ruled out as contrary to the observed action of the universe in opposing momentum changes (proportional only to the mass of the accelerated object and not Hubble mass as a whole). M_H is too large by a factor of $(10^{26})^2$ whereas volumetric density ρ_U is too small by a factor 10^{-26}. As an organic composite, the Hubble functions neither as a vacuum nor a solid, but as an apparitional area density.

The tie between *mass as inertia*, *space as area* and *acceleration as gravity*, is to be found in the means by which the universe exerts dominion over its parts. Reactionary force(s) arise, not from space-mass amalgamations constituted as a volume, but as a plurality of parallel planes. It is by this topology analogy, the universe marshals a bastian of simultaneously concurrent forces. The reactionary force exerted upon an accelerating body by an infinite plane, is always perpendicular thereto. But inertia, unlike gravity, has no extended field beyond the surface of the plane itself. How then do distant planes interact to oppose the acceleration of a local mass?

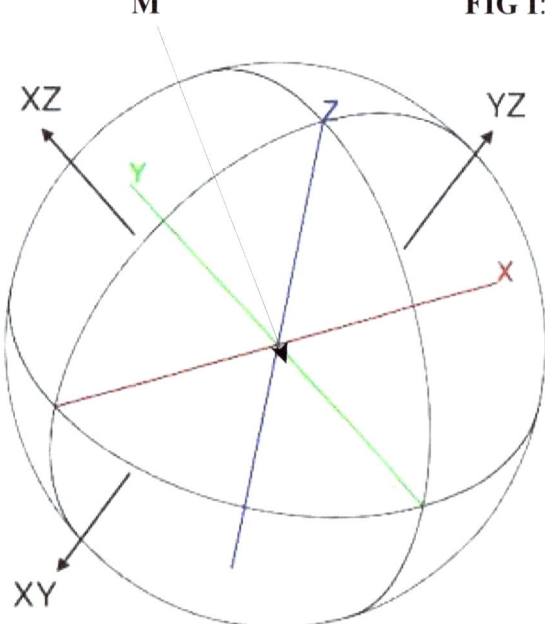

FIG I: A mass **M** moving with uniform velocity '**v**' views the Hubble as a concentric sphere. The Hubble content is indifferent to non-accelerating objects, offering neither opposition nor evidence of motion with respect thereto. Yet a slight change in speed or direction of **M** provokes immediate response from the universe to offset momentum change. Understanding space as an operative inertial area density, requires a simultaneously reckoning of gravity as a consequence of inertial reaction imposed by spatial expansion.

Fields created by flat planes are normal thereto, hence force-lines emanating therefrom, are parallel, and consequently intensity does not diminish with distance. A single area-density plane can be divided into a plurality of lesser area-density planes. There is thus no difference between the field created by a single plane and the field created by slicing the single plane into many parallel planes. Conversely, a plurality of parallel planes can be considered gravitationally equivalent to a single plane.

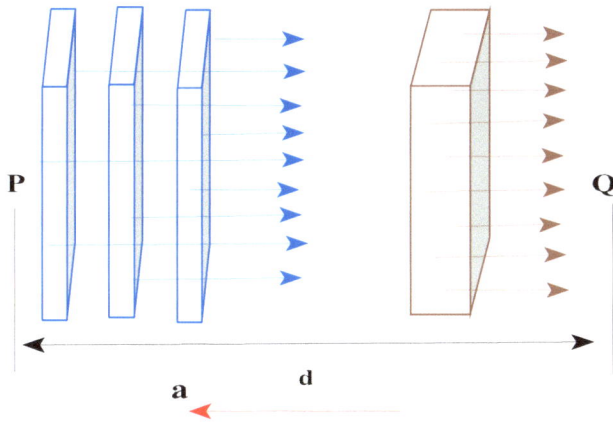

Fig II: Four accelerated Planes

All planes are accelerated at a rate 'a' (in the direction of the red arrow). Each blue plane has an energy density σ_b equal 1/3 the brown plane $3\sigma_b$. At any distance 'd' measured from point P, total reactionary force intensity is $a[6\sigma_b]$ which is felt only internally by each element according to its mass. If all planes are of the same uniform density, the 3 blue planes collectively experience the same reactionary field as the brown plane. Assume the planes at rest and space (the universe) accelerated opposite to the red arrow. The artificial 'g' field intensity at distance 'd' (the parallel spatial field) is independent of distance. What is observed, for directionally accelerated masses, the universe responds as an anti-parallel counter field. Such fields are only known to be produced by flat planes.

Fig III illustrate how the universe can be modeled to exhibit the magical properties of infinite planes within the confines of the finite Hubble sphere. Five slices [3-8] divide the volume into seven parallel slabs having approximately equal matter content. From the 'g' perspective of a mass **M** at the Hubble center, each slab has approximately the same density and consequently each contributes roughly the same field intensity to the inertial area density at the Hubble center. It is by this pertinence, the universe marshals matter content to oppose acceleration. The reality of the Hubble as a laminar cannot be avoided nor can it be transformed away. In a geometrically flat universe, the number of slabs into which the volume can be sliced, is unlimited. To oppose acceleration to the degree required by Newton's 2nd law, the Hubble must act collectively as a unified plenum. Inertial-ly, only 1/3 of the Hubble mass is available to define area-density opposition to a unidirectional acceleration (inertial impedance being equally divided between the three dimensions of space (the three mutually perpendicular area density faces share M_H equally). To estimate the inertial density of a single plane in any direction orthogonal to the trajectory of an accelerating body, divide the estimated Hubble Mass by three and spread it over a plane having an area equal to that of a great circle drawn through the Hubble center.

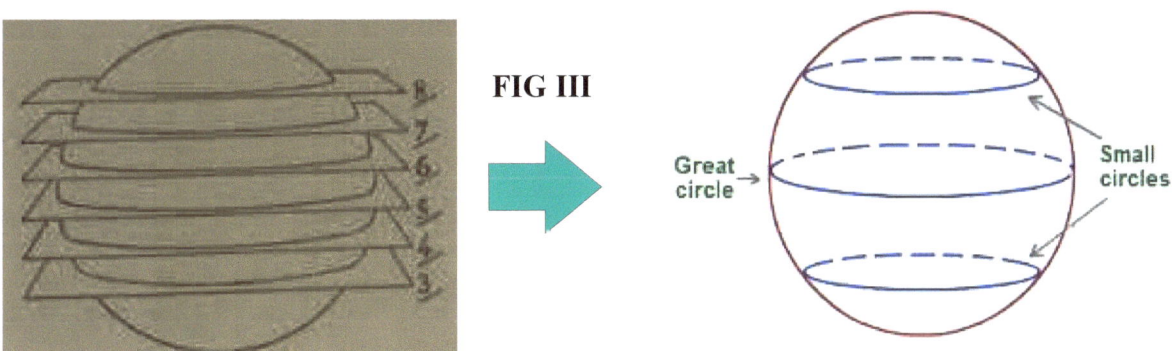

FIG III

For a Hubble constant **H** = 70, the radial scale R_H of the Hubble sphere $\approx 1.3 \times 10^{26}$ meters, then

$$\sigma_H = \frac{M_H}{\pi R_H^2} = \frac{[1.5 \times 10^{53}\,\text{kg}]/3}{(3.14)(1.3 \times 10^{26}\,\text{meters})^2} = 0.94\,\text{kg/m}^2 \tag{P-1}$$

Fig IV: An alternative to the great circle as an area for settling Hubble mass. In this rendition, a cubical volume having a side length "**L**" is plucked from the Hubble to construct three orthogonal planes (**Fig V**). Hubble mass is divided into three separate directional densities. A cube of side "**L**" has volume L^3 and three orthogonal surfaces areas L^2, so

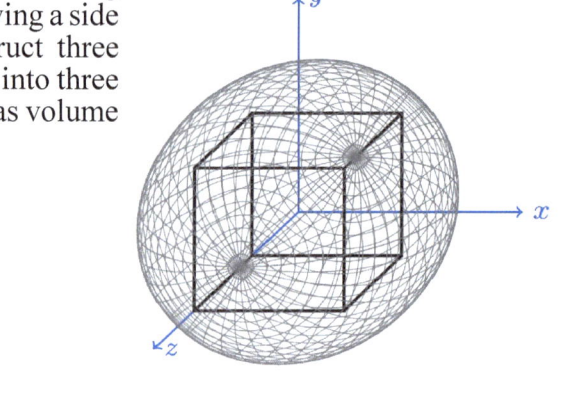

FIG V

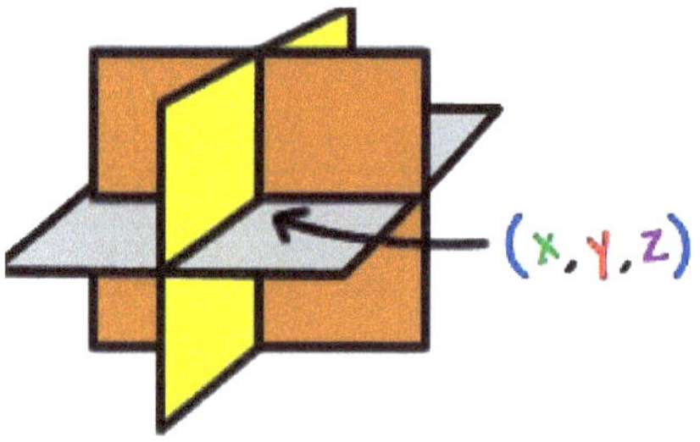

FIG IV

$$M_H = \rho_H(L^3) \quad \text{and} \quad M_H = \sigma_H(3L^2), \quad \text{Whence} \quad \sigma_H = \rho_H(L^3)/(3L^2) = \rho_H(L/3) \quad \text{(P-2)}$$

FIG VI: That the universe can act virtually as a set of orthogonal planes (**Fig V**), it can also be modeled as a set of parallel planes (each set orthogonal to the other two). The operative essence of space as dynamic reactance to accelerated motion, does not follow from its ostensible form as a vacuous volume lightly sprinkled with matter. The same agglomerative perceived as a plurality of parallel area densities gives the universe an intelligible physical coherence as an isotopic area-density To extend **Fig VI** to a panoptic, the volume of the cosmic sample must be strati graphed as a plenum of planes. Reckoned from the perspective of an accelerating mass, the emulation can be reduced to a single plane which contains the entire Hubble mass. As above belabored, a single high area-density plane is fully equivalent to a plenum of laminae filling the same volume.

FIG VI

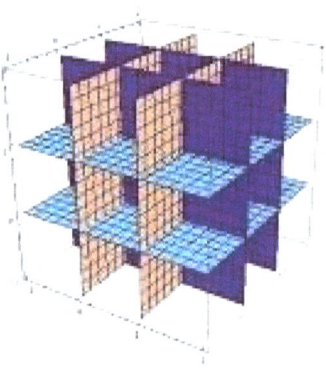

To understand inertia, one must take the road not taken by Einstein. Viewing the universe as a static 3-D volume leads to a dead end. Considered as a plenum of additive densities, however, the Hubble universe reduces to a single isotopic area density σ_H. Inertial response is instantaneous because reactive force is intrinsically superimposed upon an accelerating body at its immediate location within the area density defining composite of all planes purposed to construct the cumulative area density. Consistent with observation, inertial reaction is local. The essence of infinite plane area-density is a ubiquitous and omnipresent virtual property of the collective effect of mass in all planes participating in the construct of the composite. Mach's Principle is recovered fully extant[1]

---

[1] The Nineteenth Century physicist, Ernst Mach, proposed "Inertial Forces" to be the result of other matter rather than the constitutional endowments of empty space. Einstein was influenced by this idea, and initially attempted to incorporate, Mach's Principle, as a rudiment of General Relativity. He later changed his conviction, erroneously because it appeared to require instantaneous action-at-a-distance). Both Mach and Einstein, however, rejected the idea of inertial reactance as being solely an internal characteristic of mass (unrelated to the universe). Neither, however, progressed much beyond conjecture. To admit a cosmological source within the predicate of Mach's Principle, the cosmological contribution must be unity. Both (P-1) and (P-2) are of order of magnitude '1'

While there is no mathematical formalism for directly converting volumetric density to flat plane area density, [as estimated by (P-1) and P-2)], there is a Theorem due to Gauss which mathematically transforms volume integrals to surface integrals. If the average Hubble density is denoted by ρ_H, then the transformed area density σ_s can be expressed as:

$$\int_V \rho_H \mathbf{dv} = \int_s \sigma_s \, \mathbf{dA} \tag{P-3}$$

where ρ_H is the average density of the Hubble sphere and σ_s is the average density of a 2-sphere shell having the same mass and size. For the Hubble as a 2-sphere, the integral over the area is $4\pi(R_H)^2$ and as a 3-sphere, the integral over the volume is $[4/3][\pi(R_H)^3]$. Hence:

$$\sigma_s = \frac{R\rho_H}{3} \tag{P-4}$$

From the earth, the universe appears as an expanding concentric sphere, but as a surface density, (P-4) has the same form as (P-2). Transformation of the Hubble mass to a far away surface-density, leaves the earth devoid of tactility therewith. But our Hubble center is at once located at the edge of many Hubble spheres centered upon our own Hubble sphere (**Fig XII-A**). Just as the mass of the Hubble sphere centered on the earth can be transformed to an effectual surface area-density of our own Hubble sphere, so also other Hubble spheres may be likewise transformed. In particular, as shown in **Fig VII-B**, an accelerating local mass **M**, (red arrow) experiences the combined affect of the two Hubble area-densities superimposed upon **M** at their tangent point. Since each of the two transformed spheres is centered on the line of action coincident with the acceleration, **M** experiences cosmic area-density twice over, once from the operative area density of the shell ahead, once from the operative area density of the shell behind. There is nothing special about the earth as a preferred location, ergo space exists as a double Hubble area-density (dotted red plane) everywhere. The plane however, serves double duty -- as an impedance in each direction, thus reducing effective opposition to unidirectional acceleration by 50%. Both shell densities, however, require adjustment to account for lost gravitational energy in transforming from 3-sphere to 2-sphere.

1. Energy lost in transforming volumetric density ρ_H to area-density σ_H. Energy difference due to gravitation mass deficit between 2-sphere and 3-sphere is reduced by a factor 5/6.

2. Energy lost in transforming a spherical surface density to an infinite flat plane results in a 50% loss of gravitational energy.

3 Energy gained by the superposition of two area densities at all points of coincidence Increases overall density by a factor of 2.

Inasmuch as the effect of the 2nd and 3rd factors cancel, net area-density will be determined solely by Factor #1. Transformation from 3-sphere volumetric density to 2-sphere area density is not the same as building a 2-sphere from the same mass **M**. However, in order for the universe to oppose acceleration in accordance with Newton's 2nd Law, it must function as a virtual area density everywhere. As a virtual area-density, the cosmos cannot claim existence inconsistent therewith, that is, it cannot have the same mass and radius of a 3-sphere when it functions as a 2-sphere. (Once disassembled into bare mass and reconstituted as a 2-sphere, 3-sphere gravitational energy is lost).

FIG XII-A   **FIG VII-B**

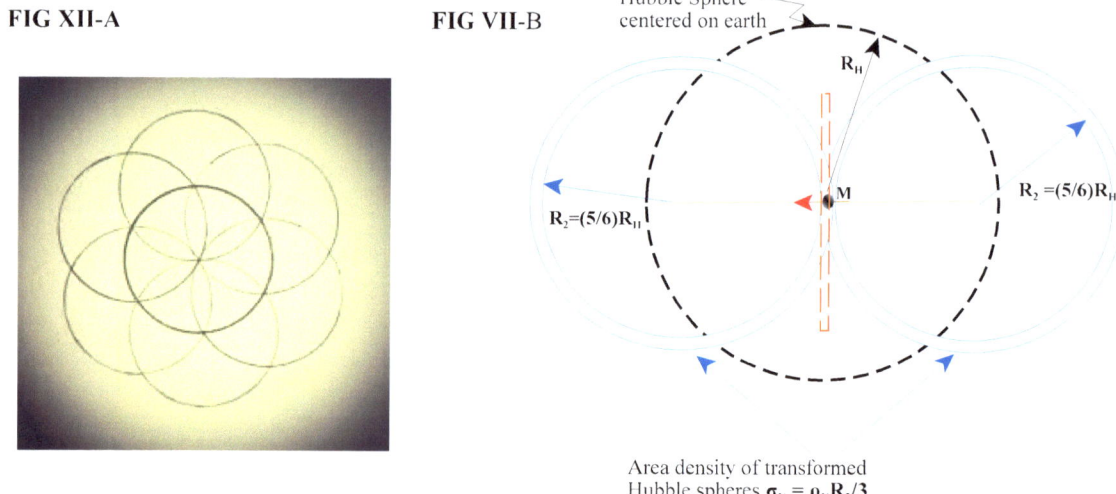

Area density of transformed Hubble spheres $\sigma_t = \rho_t R_2/3$

Fig VII-A: Every point is the center of its own Hubble sphere (bold black), and at once the locos of all Hubble spheres centered on the manifold defined thereby. **Fig VII-B**: An accelerating mass **M** experiences the universe as a virtual area-density everywhere (dotted orange plane). Every accelerating mass is instantly opposed by metaphorical superposition of two such Hubble spheres (blue) centered on a line of action perpendicular to the acceleration (red). From the perspective of **M**, they combine to form an orthogonal plane (dotted orange), the strength of which is revealed by superposition of the two surfaces each created by transformation from 3-sphere volume density to 2-sphere area-density.

As corporal 3-spheres, each has positive bare mass energy $M_b c^2$ balanced by a 'g' field sphere of negative energy $U_3 = 3G(M_b)^2/5R_3$. As operative virtual area densities, each shell will contain the same bare mass factor, but distributed over a smaller area to account for the loss of gravitational energy attributable to 2-sphere geometry.[2] As operative 2-sphere area-densities, each transformed Hubble sphere retains the same bare mass energy $M_b c^2$ and exhibits an equal negative gravitational energy $U_2 = G(M_b)^2/2R_2$ equal to U_3 when $R_2 = (5/6)R_3$, that is[3]

$$M_3 = M_b + U_3 = 2U_3 = \frac{6(M_b)^2 G}{5R} \tag{P-5}$$

Energy of a 2-sphere is:

$$U_2 = \frac{(M_b)^2 G}{2(R_2)^2} \tag{P-6}$$

---

[2] The efficacy of inertia as area density does not involve a physical location per se -- nature displays opposition to acceleration consequent to the requirement, momentum and net energy, be zero on the global scale. This self balancing condition requires the universe function as a ubiquitous area-density rather than volume density. Experiments reveal the apparition of the universe as $=1/kg/m^2$ area density. The visual impression of the universe as a 3-d Volume obscures the subliminal laminar structure. It is the underlying functionality of the universe as an area-density that explains Newton's 2nd Law. What is measured by inertial reaction, is the existence of an operative form that is vastly different than its appearance.

[3] Bare mass is the sum of the masses measured as many small pieces sufficiently separated to avoid mensurable gravitational interaction

Transformation of Hubble Mass-Energy to Infinite Plane

Both Newton and Einstein opined that inertia could not be an antonymous property of the body undergoing acceleration, independent of the universe. But to admit a cosmological construct within the auspicious of Mach's Principle, was to Einstein, an effrontery to Special Relativity. Herein, the mystery of inertial space as instantaneous communicator of distant matter is addressed by double transformation of two Hubble 3-spheres to a metaphorical infinite plane as depicted in **Fig XII**. If the two Hubble spheres have radii R_3 commensurate with the Hubble sphere centered on the earth, transformation of inertial mass to an operative 2-sphere surface having the same area would reduce energy by a factor of **5/6**. To recover the gravitational energy lost in reconfiguring 3-spheres as 2-spheres using the same bare M_b, the 2-sphere radius R_2 is shrunk by a factor **5/6**. Transformation of the two 2-sphere unison at **P** then merges to a plane, taking into account the gravitational energy deficit for each sphere as 50%. For present purposes, R_3 will be provisionally taken as equal to the present estimate of the Hubble scale $R_H = 1.3 \times 10^{26}$ **meters** based upon $H = 70$. As previously, cosmic bare mass M_b is assumed to be in the range $\approx 1.5 \times 10^{53}$ **kgm**, the effective radius R_2 for purposes of calculating infinite plane area density is then $(5/6)(1.3 \times 10^{26}) \approx 1.1 \times 10^{26}$ **meters**.

| Uniform Density Hubble 3-Sphere with G energy $U_3 = [3(M_u)^2][G/5R_3]$ | Transformation of 3-D sphere to 2-sphere with no energy change reduces R_2 | 2-sphere transformation to infinite flat plane reduces gravitational energy 50% |

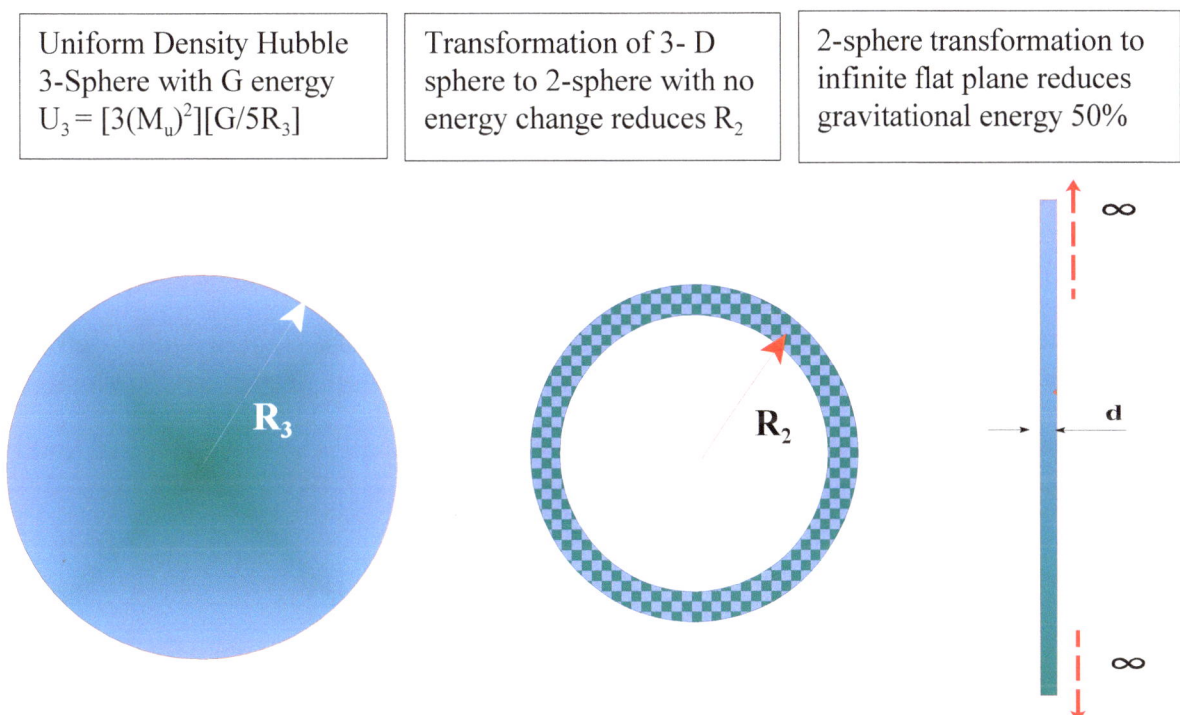

Fig XIII Physically, the Hubble is 3-D, but operatively it is a surface density σ_U rather than a volume density ρ_U. Intuitively, one might reason the transformation from 2-sphere to infinite plane could be carried out by simply setting $R_2 \rightarrow \infty$ (Assuming new volume adds mass proportionately, so density remains constant). But alas, that is not the way the world is made -- transformation from a spherical surface-density to a flat plane having the same area, halves the effective area-density.[4]

---

[4] The infinite 2-sphere and infinite plane are mathematically different geometries. Half of 2-sphere energy is in the form of gravitational binding, which is lost when transforming from 2-sphere to infinite plane.

From an inertial perspective, the universe acts as a local area-density. Gravitational energy $M_{g3} = [(3M_b^2G)/5R_3]$ is larger than area density $M_{g2} = [M_b^2G/2R_2]$. The effective radius R_2 is thus determined from (P-7):

$$U_3 - U_2 = \frac{3M_b^2G}{5R_3} - \frac{M_b^3G}{2R_2} \tag{P-7}$$

From which

$$R_2 = (5/6)R_3 \tag{P-8}$$

Equation (P-8) provides the final step in transforming nearly empty space to a form capable of explaining inertial impedance. But if the universe is to act as an area-density (as it must to explain 2nd law inertial reaction(s) in terms of infinite plane mechanics), the only energy available is that which corresponds to its operational mode. As area-densities, 2 spheres have less gravitational energy, but both the 2-sphere and 3 sphere contain the same bare mass. Because 2-spheres transformed from 3-spheres have the same bare mass as the 3-sphere from whence derived, the operative value of both R_2 spheres must be reduced to $(5/6)R_3$ per (P-8). The total energy of the Hubble universe as a 2-sphere operative area is the therefore:

$$E_T = \frac{M_b c^2}{1} + \frac{M_b^2 G}{2R_2} = \frac{M_b c^2}{1} + \frac{M_b c^2}{4\pi R^2{}_2 \sigma_U} \tag{P-9}$$

At this juncture, the 2-sphere view of the Hubble sphere as an interim configuration state, would be characterized as having a total energy:

$$E_T = Mc^2 + U_2 = Mc^2 + \frac{M_b^2 G}{2R} \tag{P-10}$$

In a zero energy universe, positive matter energy Mc^2 equals negative gravitation energy U_2, in which case, one might erroneously assume an infinite radius spherical shell would produce the same result as an infinite area flat plane having the same area density. But a shell universe has zero internal gravitational force whereas the infinite flat plane is deemed to provoke an equal uniform gravitational force in opposite directions normal to the plane. For present purposes, gravitational implications can be disregarded, whence the operative mass and effective area reduce to:

$$\sigma_U = \frac{M_b}{4\pi R^2} \approx \frac{1.5 \times 10^{53}\,kg}{(12..56)(1.1 \times 10^{26}\,meters)^2} \approx \frac{one\ kg}{meter^2} \tag{P-11}$$

For the values selected, notice is taken of the fact area-density σ_U is approximately unity in mks units. That it must be exactly "*one kg/meter²*" when scaled using Newton's 2nd law, will revitalize with compelling cogency, the long prophesied dependence of local inertia upon global mass within the predication of Mach's Principle.[5] For the present, the question as to whether the universe can be prospected as an area-density, has been answered in the affirmative.

---

[5]Pursuant to the findings revealed herein, the debarment of Mach's Principle will be lifted. As a volume thinly populated with chunks of matter, the Hubble plays no part in the creation of inertial reaction. Cognition of the cosmos as a composite laminar, renders the void apprehensible. The smoothness and isotropy of the CBR testify to the influence of all matter upon all other matter. The puzzlement(s) of the horizon needs reconsideration.

To explore σ_U as a virtual inertial area-density throughout, each infinitesimal mass within an accelerating body must separately interact therewith —> as a free body **B** of mass M_B having an effective cross sectional 'A' with respect thereto. Because inertial action of the infinite plane can only involve those lines of action normal to the surface which also conjoin M_B, the interaction between the universe **U** and **B** will be confined to the common area, i.e., the projection of **A** upon **U**. Since σ_U is the infinite plane emulation of cosmic mass in form as area density, the imaginary acceleration opposing force lines will be perpendicular to σ_U. In communicating σ_U as area-density impedance, it is convenient to express M_B as area-density σ_B.[6]

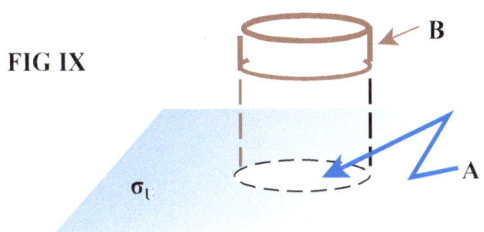

FIG IX

Fig IX illustrates the mutual domain of action between **U** and **B** as the cookie cutter area 'A' punched out by the projection of **B** thereon. The void (represented as the area density σ_U), appears as a coincident co-planer superimposed area orthogonal to the acceleration (which can be either up or down wrt σ_U inasmuch as spatial impedance is bidirectional and therefor acts to oppose an increase or decrease in **B**'s velocity

Applying Newton's 2nd law to the segment defined by the projection of **B** upon σ_U, the normalized expression for reactionary force per unit area in terms of the mass area density of **B** is:

$$\frac{F}{m^2} = \frac{M_B}{m^2} \mathbf{a} \qquad (P\text{-}12)$$

Force **F**, however, can be expressed as cosmic [mass/area] multiplied by $\mathbf{a}_2$ with respect to **B**:

$$\frac{M_U}{m^2} \mathbf{a}_2 = \frac{M_B}{m^2} \mathbf{a}_1 \qquad (P\text{-}13)$$

Einstein's doctrine of relative acceleration teaches, there is no difference whether **B** accelerates with respect to the universe or the universe accelerates with respect to **B**. By this reciprocity, spatial acceleration endowed with the characteristic area-density σ_U. An accelerated mass **B** encounters σ_U as pressure created momentum flow $[(\sigma_B)\mathbf{a}_1]$.[7] Expressing (P-13) in terms of σ_U and σ_B:

$$\frac{F}{m^2} = \mathbf{a}_2 \sigma_U = \mathbf{a}_1 \sigma_B \qquad (P\text{-}14)$$

---

[6]That any irregular shape, size or non-uniform density can be transformed into an idealized uniform flat plane area density for purposes of calculating the interaction of the universe thereon, will be recognized as one more adjunct of that property of infinite planes which makes action at a distance independent of the distance.

[7]Likewise, expansion created accelerations (in form as 3-D isotropic spatial recessional flow), create momentum pressure $[(\sigma_U)\mathbf{a}_2]$. As depicted in **Fig VII**, both the universe and **B** are projected as area densities, expansion is isotropic, so force lines can be radial rather than parallel when the acceleration is isotropic.

From (P-14), an accelerated mass **M** having cross sectional area '**A**' normal to the direction of acceleration '**a**' will be opposed by the universe with force **F** given by:

$$F = Ma = \sigma_U a[A] \tag{P-15}$$

For **one kg** of mass formed into a uniform density having area '**A**' equal **1 meter²**, the reactionary force **F** = **one ntn** when accelerated at **a** = **one meter/sec²**. That reactionary force is independent of orientation relative to the direction of acceleration, is testament to the continuity and uniformity of σ_U. Area-density is the subliminal form of cosmic opposition to acceleration, it renders unto space, the status of isotropic continuum. As first espied by Newton, and later phrased by Feynman, the public feature of physical law rests with the spatial frame into which the physics is put.

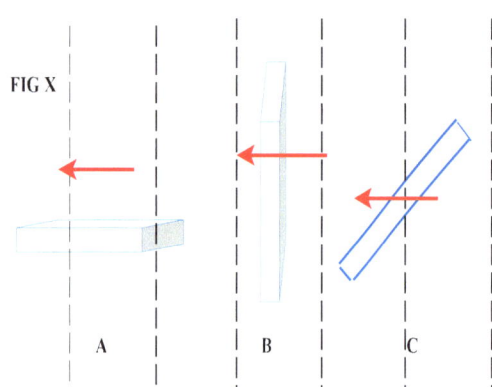

Fig X: The dotted vertical lines (black) representative of σ_U constitute a continuous laminate of infinitely thin contiguous laminae. Area-density vitalized by acceleration translates to momentum flow (pressure). To evaluate force/area, it is necessary to know the interaction area '*A*' as covered in the discourse of **IX**. While the mass of **B** can be spread over any area '*A*' for purpose of determining pressure created by a given acceleration applied thereto, there is a common area '*A*' that represents the cookie cutter area of action between the universe and an accelerating mass, which cancels on both sides of the equation leaving the familiar form of Newton's 2nd Law.[8]

To calibrate area-density, '**a**' will be taken as [**one m/sec²**] and M_B as one **kg**. Reactionary force is:

$$F = M_B a = a\sigma_B A = one \ \ ntn \tag{P-15-A}$$

Since the acceleration of M_B wrt the universe {U} is equal to the acceleration {U} wrt to M_B, then the reaction pressure created by the universe is $(a)(\sigma_U)$

$$P_U = F_U/m^2 = 1 \ ntn/m^2 = (a)(\sigma_U) \tag{P-16}$$

Using the orientation "**B**" in figure **Fig X** (normal to acceleration) as a dummy plane having area density equal to what is believe to be approximately **1 ntn/m²**, then:

$$\sigma_U = \frac{1 \ ntn}{aA} = \frac{\dfrac{1 \ kgm \ meter}{sec^2}}{\dfrac{1 \ meter}{sec^2}\left[1 \ meter^2\right]} = \frac{1 \ kg}{meter^2} \tag{P-17}$$

---

[8] Reduction of an expression to its simplest form can sometimes result in a loss of critical information necessary to explain the result. While Newton's 2nd law was experimentally formulated, the final form gives no indication of the underlying area dependent factor, leaving thereby the false impression of a universe magically endowed with the power to sift out mass to the exclusion of its area density relationship to the cosmos.

Inertial Space as a Continuum

A different orientation would result in a different common area of action - but no change in reactive force (as is well known, inertial reaction is area independent). What is common for all orientations, configurations and geometries, is the ubiquitous presence of the orthogonal σ_U area density at every locality. As shown in **Fig XI**, each element of an accelerated body **B** simultaneous feels the universe as momentum flow (the pressure **P** created by *"acceleration x area density"*). But **P** is unlike the pressure felt by a kinetically bombarded surface. Spatial inertia is a condition of the universe, there are no force conveying particles or momentum transferring quantum(s) to be found and none need be invented. The only motes involved in the inertial process are masses in the form of particles and energies that bind them together. In **Fig XI**, the particles are depicted as 3-D spatial vortices (blue), (a construct for justifying the mass energy of subatomic particles). As such, 3-D circulatory inertial space exhibits inertia as mc^2 energy subsumed within the concept of circulatory space. There is not a particle in the universe that can be identified as static mass. The eleven rotational energies making up body **B** together with binding energy(not shown), constitute the inertial mass-energy thereof. Just as M_B contributes to the total mass of the universe, the composite area-density field σ_U is reflected back upon 'B as a cosmic counter field of all individual masses contributing thereto. Each particle thus carries its own coincident tie to the whole as represented by the dotted vertical lines (metaphorical virtual mini planes indicating the ubiquitous presence of the area-density function σ_U) each orthogonal to the acceleration of '**B**' and coincident therewith.

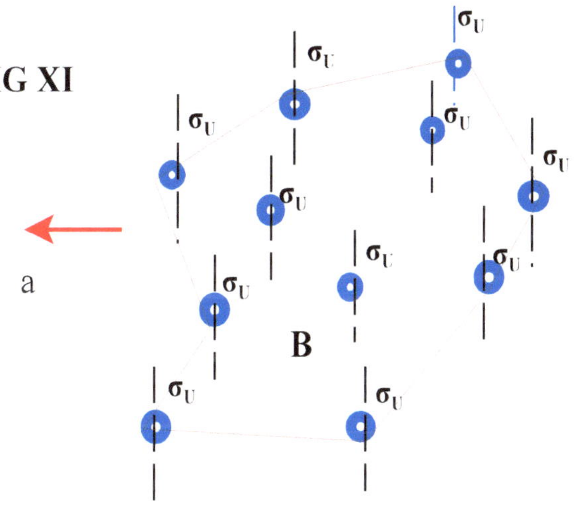

Fig XI: An accelerated body **B** comprising 11 particles held together by electrical and quantum forces. Non accelerating masses experience the universe as a near perfect vacuum save for the influences of thinly scattered matter throughout the void. All accelerating masses, however, experience the universe as a superimposed virtual area-density σ_U. Each particle thereof instantly made aware of an alternative factual state of the universe as a virtual plenum of area-density planes presenting an inertial face to acceleration from cosmic mass near and far.

From the mystery ratio,[9] the square of the velocity of a photon escaping from the universe c^2 equals $[M_b G/R_2]$ where R_2 is the effective Hubble radius 1.1×10^{26} meters and M_b is the bare mass (1.5×10^{53} kg). The '**g**' field of the Hubble based upon R_2 and $M_b = [M_b G/R_2^2] = c^2/R_2$. Transformation of M_b to the effective area of the Hubble surface $[4\pi R_2^2]$ refashions the Hubble as a 2-sphere area density operative

$$\sigma_U = \frac{M_U}{4\pi R_2^2} = \frac{1.5 \times 10^{53} \text{ kg}}{12.56 \times (1.1 \times 10^{26} \text{meters})^2} \approx \frac{1 \text{ kg}}{\text{meter}^2} \qquad \text{(P-18)}$$

---

[9] Cosmologists have been long perplexed as to why $M_U G/Rc^2 = 1$ within the limits of experimental error. Why should cosmic mass M_U multiplied by **G** = Hubble scale **R** multiplied by c^2. In their efforts to relate inertia to the idea that each particle feels the presence of all other particles in the universe (Mach's Principle), Robert Dicke and Carl Brans, proposed $M_U G/Rc^2 = 1$, be interpreted as the ratio of gravitational mass to inertial mass. It should be now known that the mystery ratio is but an expression for **G** = [Hubble volumetric expansion $c^2 R$ /Hubble Mass M_U]

Emergence of Gravity

Adverting again to **Fig XI**, the question arises as to the nature of the inertial force, if any, that arises if **B** is at rest within the spatial environment of an accelerating universe. Einstein pondered the problem while working out the Theory of General Relativity, concluding that a stationary object would feel a unidirectional boost of the universe as an inertial reaction. As far as is known, Einstein did not at any time, before or after the discovery of expansion, consider the effect of inertial mass upon isotropic spatial expansion. To Einstein, acceleration like velocity, only had relational meaning. For whatever reason (which he did not elaborate), the universe in some fashion, was deemed to convey changing momentum throughout the distend of space.

Boosting the cosmos is not an experiment that can be performed. Conveniently, a continuous ready made isotropic acceleration field exists in form as spatial expansion." It would be axiomatic that any reactionary force attributed thereto, also be isotropic. For a test mass of known density, any body in the solar system will due. The earth being most accessible, it will be taken as the beta test site for determine the strength if the reactionary field created by the cosmological expansion field?

For a spherically uniform body undergoing unidirectional acceleration '**a**,' the counter pressure is determined by the density σ_U of the infinite plane multiplied by **a**. The pressure $\sigma_U(a)$ thus defines momentum flow influx. For an isotropic acceleration field acting upon a body **B** composed of bound particles (not expanding) the reactionary force will be opposite to the direction of spatial acceleration, that is, inwardly directed normal to the surface of **B**. As is the case when **B** is accelerated with respect to space, the reactionary force created by expanding space must be supplied by the σ_U density field that represents the universe in all inertial matters.

Inertial matter subjected to isotropic acceleration supplied by spatial expansion is on a different footing than unidirectionally accelerated mass. Conservation of momentum demands every force be balanced to zero. Reactance pressure created by the action of spatial acceleration upon inertial matter must be balanced by negative pressure -- momentum flow inward (convergent upon the surface '**S**' of our spherically uniform density). Thus if total force **F** integrated over the surface area is A_B is:

$$\mathbf{F} = \mathbf{M_B}(\mathbf{a_n}) \tag{P-19}$$

Inertial pressure created by exponential expansion acting upon the shell density σ_B is

$$\mathbf{P_I} = \sigma_B(\mathbf{a_n}) \tag{P-20}$$

Reactive pressure by the universe is: $\quad \mathbf{P_R} = \sigma_U(\mathbf{g}) \tag{P-21}$

Approximating earth as a uniform area density σ_B, then for a cosmological acceleration factor:

$$\mathbf{a_n} = \mathbf{c}^2/\mathbf{R}, \text{ then:}$$

$$\mathbf{g} = \frac{\sigma_B}{\sigma_u}\left[\frac{\mathbf{c}^2}{\mathbf{R}}\right] = \frac{\mathbf{M_B}\mathbf{c}^2}{\sigma_U 4\pi R(r)^2} = \frac{\mathbf{M_B}}{\mathbf{r}^2}\left[\frac{\mathbf{c}^2}{4\pi R(\sigma_U)}\right] \tag{P-22}$$

And from Newton's law of Gravity

$$\mathbf{g} = \mathbf{M_B} G/\mathbf{r}^2 \tag{P-23}$$

Then from (P-22)

$$G = \frac{\mathbf{c}^2}{4\pi \mathbf{R} \sigma_U} \tag{P-24}$$

Gravity As Pressure

From (P-22) and (P-24), together with the depiction of earth and universe configured as area-densities per **Fig XII,** then:

$$M_e = 5.98 \times 10^{24} \text{ kgm}$$
$$r_e = 6.37 \times 10^6 \text{ meters}.$$

The surface density σ_E of earth in planer form is:

$$\sigma_E = \frac{M_e}{4\pi(r_e)^2} = \frac{5.98 \times 10^{24}}{(12.56)(6.37 \times 10^6)^2} = \frac{5.98}{509.65} = 1.173 \times 10^{10} \text{ kgm/meter}^2 \qquad \text{(P-25)}$$

Expressing earth as a non expanding surface density experiencing the constant stress of ongoing spatial divergence, one arrives at a poor-mans explanation of gravity as surface pressure.

$$P_E = \frac{\text{momentum transfer rate}}{\text{area}} = \frac{M}{A} \times \frac{dv}{dt} \qquad \text{(P-26)}$$

Artificially shifting all interior mass to the earth's surface, creates the erroneous impression that the earth's '**g**' field can be likened to surface pressure akin to the atmosphere. The reality is, the product [*earth's mass multiplied by the spatial expansion factor c^2/R*] yields total force as if earth's mass were concentrated on its surface. When total force is divided by surface area, the dimensionality has units (*ntn per square meter*). There is, however, no pressure per se. What is revealed by the units is actually momentum flow into the earth. So while false pressure numerically equals rate of momentum flow at the surface, flow continues at a diminishing rate throughout the interior.

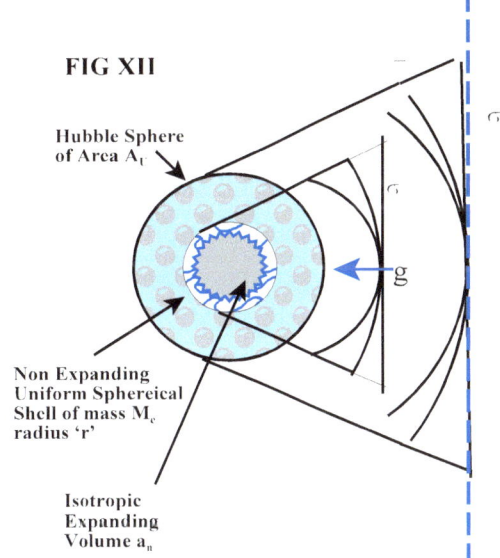

What is missing from (P-24) is the global reaction of space to the distortion of the expansion field created by the earth's mass. Counter pressure is the result of momentum influx across the earths surface. Equating cosmological pressure with the reactionary pressure created by (P-24), there results:

$$(\sigma_u)a_2 = \sigma_E a_n \qquad \text{(P-27)}$$
$$a_2(\sigma_U) = a_h(\sigma_E)$$

Since a_2 is simply the reactionary acceleration '**g**,' then from (P-21):

$$g = \frac{\sigma_e}{\sigma_U} a_n = \frac{1.173 \times 10^{10} \frac{\text{kgm}}{\text{meter}^2}}{\frac{\text{one kgm}}{\text{meter}^2}} \times \frac{9 \times 10^{16} \left[\frac{\text{meters}}{\text{sec}}\right]^2}{1.1 \times 10^{26} \text{ meters}} = 9.6 \text{ meters/sec}^2 \qquad \text{(P-28)}$$

There are thus two pressures in balance. Solving for '**g**' thus outputs the earth's '**g**' in conventional units of acceleration (meters/sec^2)

Second Law Symmetry

While the empirical evidence for σ_U as **one kg/m²** has been consistent and persuasive, in the last analysis, all must conform with the physical principles upon which physics is founded.[10] Any theory of gravity dependence upon inertia must correspond to Newton's law of Inertia.[11]

By convention, a mass **M** accelerated at [**one meter per second squared**] provokes a cosmic responsive force of [**one 'ntn'** per '**kg**']. But was it always so? What has been long known, is that, in addition to exhibiting an inertial reactance proportional to **M**, there is also an accompanying '**g**' field, also proportion to **M**. And while the '**g**' field of **M** presumably extends to the limit of the universe, the inertial effect of space upon accelerated mass is not explained by standard theory. Inertia appears to be something local, an intrinsic property of mass itself, unrelated to the universe? Yet appearance is frequently beguiling The '**g**' field of **M** that depends upon the inertial property **M**, is an infinite distension thereof?

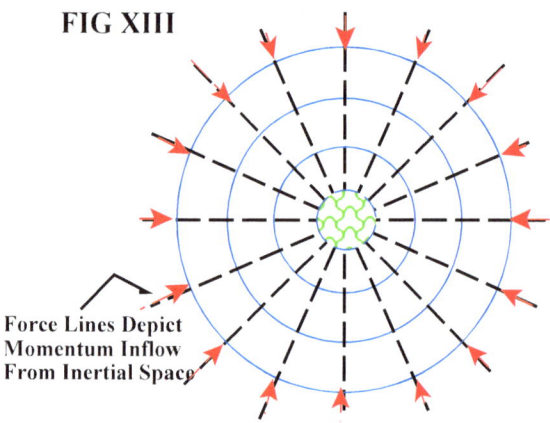

FIG XIII

Force Lines Depict Momentum Inflow From Inertial Space

As illustrated in **Fig XIII**, the intensity of the '**g**' field falls off inverse squared with distance, but the cumulative effect of the gravitational intensity of a spherically uniform mass, is independent of distance. The integral of the '**g**' force over an encompassing surface, *a la* Gauss, is the same at any distance. In this sense, the effect of **M** is projected equally upon any other mass fashioned as a uniform shell encompassing **M**. Hence then, perhaps the reality of the Hubble can again be exploited as a 2- sphere?

As previously developed, the fact that inertial reactionary force is equal and opposite to acceleration, strongly suggests something in the nature of a flat acceleration responsive membrane (e.g., a dv/dt) capable of reversing and rebounding a primary intrusion along the same path with the same force). Contrary to elasticity in the ordinary sense, inertia acts only upon acceleration. Masses with constant velocity are unaffected. The objectification of spatial inertia as a ubiquitous elastic sponge comports with a world where force is defined by the dynamics of change. A one kg mass spread uniformly over one square meter, creates an area-density of **one kg/m²**. Irrespective of orientation, an acceleration of **one m/sec²** provokes a cosmic reactionary force of one **ntn**.

---

[10] All genesis theories begin with fabricated initial conditions. Herein however, in lieu of all-at-once-matter-conjured-creation, the inertial property of particles is deemed to accrue proportionately with cosmic size. In this scenario, all particles are initially massless and are produced during the first instant of expansion as a result of negative pressure created thereby. The details of the transition from photons to electrons and combinations thereof, are of no moment with regard to the present thesis. However, to understand how gravity and inertia are related, one must understand how particles acquire inertia. This means looking into cosmic history. If the present rate of expansion is c^2/R, the possibility exists that it may have been infinite for an instantaneous period when $R \rightarrow 0$.

[11] *"The readiness with which a body responds to the call of an external force depends on its inertial mass." This law of Inertia, said Einstein, marks the first great advance in physics; in fact its real beginning."*

Cosmic manifestation of area density thus appears instantly at every element of every accelerating body (**Fig XI**). Minimum pressure occurs when the direction of acceleration is ⊥ to the sheet as expected for a virtual laminar construct. Thus, per Newton's 2nd Law, (P-29) will be satisfied if [?] equals one kg:

$$\frac{?}{m^2} \times \frac{1\ m}{sec^2} = \frac{1\ ntn}{m^2} \tag{P-29}$$

Consequently, the effective cosmic area density is:

$$\sigma_U = \text{one kg/meter}^2 \tag{P-30}$$

The recognition of space as a ubiquitous virtual area-density, resolves three interrelated riddle's of inertial mechanics within the framework of Mach's Principle:

1st) The effect of mass upon inertia is enhanced by a factor of one dimension (from being spread over a volume to an equivalent surface). By this intrigue, the effective impedance of space is increased from Mass divided by distance cubed to mass divided by distance squared. Inertial reaction is then consistent with its established value **one (ntn/kg)/ meter/sec^2**.

2nd) Reactive forces are instantaneous because virtual area-density is ubiquitous. That the '**g**' fields of area-densities do not diminish with distance, so also the inertial effect of virtual area-density upon the acceleration of masses will be independent thereof (Force intensity is constant, the displacement between the point of action and the virtual area density that arises to oppose the acceleration, is subsumed as a property of area density).

3rd) The specter of special relativity and the velocity of light as the limiting speed of communication imposed thereby, is abrogated by virtual locality. That which hung so heavily upon Einstein in his attempt to incorporate Mach's Principle as a rudiment of General Relativity, can now be property understood as an instantaneous inertial operative rather than a near perfect vacuum.

4th) Mach's Principle can now be fully embraced as true statement about reality. Indeed, what is now understood as the nature of space, can be celebrated as a testament of Mach's Principle. Virtual spatial inertia must be unity -- The inertial coefficient **M** cannot have a value which is different from its gravitational coefficient **M**. The effect of distant matter, being subsumed within the construct of area density, explains why there is a bit of everything in everything, while at once also explaining the action of the universe in a way that does not alter **M** as an inertia (nor as gravity **MG**, nor as momentum **Mv**, nor as kinetic energy **Mv^2/2** nor as rest energy **Mc^2**). It is axiomatic the magnitude of local inertia be insensitive to the action of global matter except as unit area-density. The influence of global matter is manifestly ineffective in modifying in the magnitudes by which force (**ntn**) and acceleration (**ntn/kg**) are commonly measured. In the words of America's preeminent theorist, John Archibald Wheeler, how could it have been otherwise?

Virtual Inertia - Spatial Impedance

Fig XIV depicts a two dimensional cross section of a flat space universe sliced into **N** virtual slabs (dotted black) which because of the cumulative nature of flat plane **g** fields, can be represented as two virtual ambient area-density planes S_1 and S_2, each representing ½ of the cosmic energy. Between the two planes, mass **M** is specified as a uniform density cube having surface area **A** per side. Each virtual slab contains a fraction **1/N** of the total cosmic energy.

Because the 'g' fields of infinite planes are perpendicular to the plane, they are parallel to each other, and therefore cumulative in effect. But the existence of the universe as a set of virtual planes only arises in relation to an accelerating mass, and then only orthogonal to the instantaneous direction of the acceleration. Otherwise the universe exists as a volume -- expansion is isotopic, gravitational intensity falls off inverse square with distance. By contrast, the instantaneous reactionary counter force of an accelerating body, is the cumulative result of every comoving elemental mass that constitutes the matter content thereof. All reactions are parallel bundles as having been retro-reflected from a flat cosmological mirror. Every acceleration (blue arrow) will have its compliment of planes reduced to a single metaphorical composite area-density imposed upon each element of mass as shown in **Fig XI**. From the perspective of an accelerating mass **M**, the ambient area densities S_1 and S_2 appear to be the source of cosmic counter pressure.

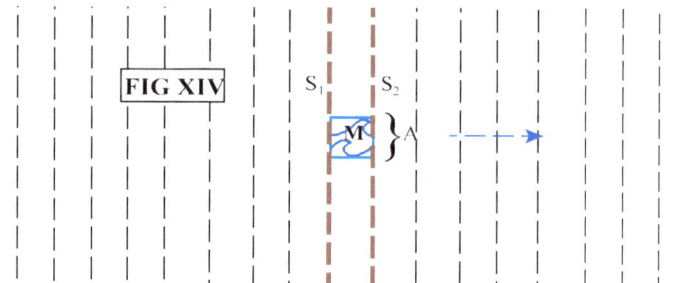

Fig XV: Green arrows represent expanding space, red arrows are momentum flow into **M** from S_2 thar arise when **M** is accelerated (blue arrow). Thus, while S_1 gets credit for representing the cosmic area-density over area '**A**,' it is the density of the individual planes that provide the cumulative inertia that determines the retro-directive reactive pressure.

S_2 itself only contains (**1/N**) the required inertial mass to explain the pressure equivalent all virtual slabs combined. While the slabs do not exist factually as separate laminae, they exist functionally as operatives. Shown by the red arrows, reactionary pressure (momentum flow) through the auspicious of S_1 acting upon **M** is directed inwardly from the universe. Newtonian 2nd law reactions, like 'g' fields, do not originate inside of masses. While the latter are 3-d convergent pressure fields, unidirectional accelerations link to the universe through parallel lines of force that connect each element of accelerating matter to the cosmos. **F = Ma** fields are continuous and parallel. Every acceleration will be greeted by the universe as a σ_U area-density (red arrows) momentum influx. Reactions are antiparallel (diametrically opposite to the direction of acceleration). Each atom, or constituent thereof, requires and receives an oppositely directed **ma** counter force, responsive thereto. The collective sum of the forces over a particular area determines the average surface pressure. Retro-directive counter action to acceleration reveals the subliminal characteristic of the cosmos as area-density.

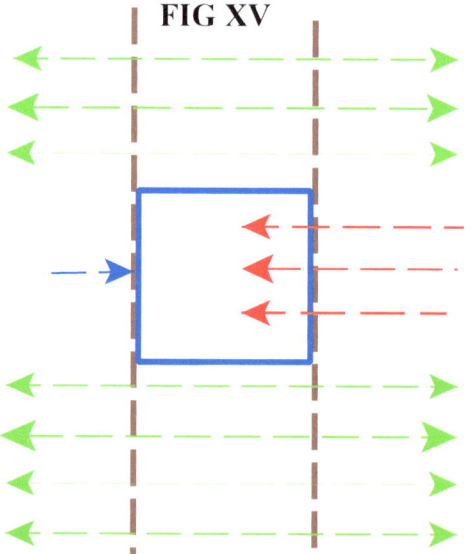

Isotropic **1/r²** distance dependent intensity applies to gravity, not to directionally accelerated bodies. The laminar model of space-mass area density comes into play for directionally accelerated objects that naturally create constant density parallel fields. They, like convergent '**g**' fields, fully account for the creation of instantaneous local forces retro directed from afar. That the number of gravitational force lines passing through all area density shells is the same at all distances, then the retro-directively converging pseudo field created by the Hubble sphere can be modeled as a single 2-sphere area-density coincident with the Hubble sphere. That cosmic mass is actually spread over a nested set of concentric area density shells, the cosmos (from gravitational perspective), can be depicted as a single Hubble surface per **Fig XVI**. Each shell has the same area-density and therefore each contributes equally to the effective inertial reactance of ambient space acting upon a local mass.

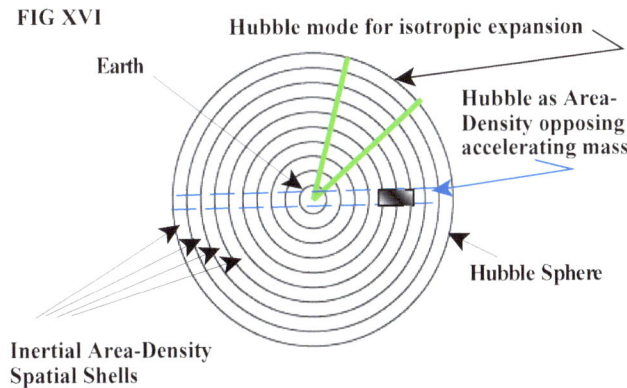

Fig XVI. In a homogenous universe, all shells will have the same area density. Spatial expansion creates inertial reactive forces, interpreted herein as convergent momentum influx called "gravity". The non expandability of matter impedes spatial expansion. While even the most dense materials are highly porous to spatial flux, matter nonetheless interferes with the free expansion of space in proportion to the displacement of the volumetric density of the mass. From Friedmann's equation:

$$G = \frac{3H^2}{4\pi\rho_U} = \frac{Rc^2}{M_b}$$

where in the first expression **G** is seen to encode the cosmological acceleration factor (**c²/R**) divided by Hubble volumetric density ρ_U. In the 2nd expression (obtained by substitution of cosmic bear mass [**M_b** divided by **V**] for ρ_U), reduces to volumertic acceleration divided by bare mass.

As example, the pressure created by expanding inertial space on a spherical body such as the earth, will be assessed by considering a functionally equivalent uniform spherical shell σ_E fashioned from a volume to surface transformation of its matter content. Likewise the operative essence of the Hubble as an impedance will be defined in terms of surface area density $\sigma_U = M_U/4\pi R^2$. The inertial relationship between the earth and the cosmos thus simplifies to **Fig XVII**. All spherical masses as well as the universe, can be considered shells for purpose of calculating '**g**' fields. However, as previously elaborated, mass deficit must be taken into account when transforming Hubble volume energy to a shell construct.

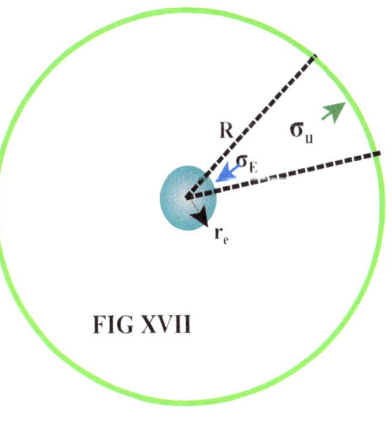

FIG XVII

As previously developed (p-28), for a cosmological acceleration factor $c^2/R = a_n$, the force created by the earth in reacting to expanding inertial space (M_E)a_n will equal the pseudo force created by the universe (M_U)**g** in reacting to the pressure sink created by the earth's mass in reactance to the spatial accerlation field created by spatial expansion.

Expressing both forces in terms of a common area, then for conservation of momentum, the two pressures will be equal, that is: The negative pressure created by cosmic expansion acting upon the inertial content (area density) of the earth will equal the negative pressure created by the reactance field of the earth acting upon the inertial content (area density) of the Hubble universe.

$$\mathbf{a}_n \sigma_E = \mathbf{g} \sigma_U$$

Spatial expansion is the sought after *"Machian connective"* between the inertia if individual bodies and the inertia of the cosmos as a whole. That expansion of free space far removed from the influence of spurious 'g' fields is isotopic, it will prove efficacious to model the universe in terms of its Hubble sphere sample size, and the parameters thereof. Inertial space then emulates as a plurality of evenly spaced area-density shells (**Fig XVI**). Unidirectional acceleration of an individual mass $\mathbf{M}_p$ experiences the universe as an orthogonal area-density corresponding to a cooking cutter section punched out of each shell (dotted blue cylinder with black oscillating piston of mass $\mathbf{M}_p$). Each small punched-out section of each spatial shell is considered a flat area density orthogonal to the back and forth motion of the Piston (**Fig XVIII**).

The reactive consequence of energy in the concentrated form of mass, is that it adversely impacts the expansion rate of space, the effect thereof for a spherically uniform mass, being greatest at the surface, diminishing inverse squared with distance measured from the center of mass. Not surprisingly, the field is commonly misunderstood as divergence emanating from within. The reality of gravity as momentum convergence reflected from the inertial reactance of other matter, follows from the unimpeachable requirement that momentum be conserved in a net zero energy universe.

Gravity does not exist as self perpetuating divergence -- cosmic participation is required to compensate for diminution in the expansion rate of space created by the impeding affect of local inertial matter. From **Fig XVII**, all such virtual flux is effectively absorbed by all other cosmic mass energy in it's operative form as the virtual area density σ_U. The resultant inertial reaction of the universe is commonly known as the 'g' field of $\mathbf{M}_E$. It is the instantaneous manifest of cosmological counteraction. It is the causal cosmological condition that brings about local 'g' fields.[12]

It is immaterial where the area-density shell σ_U is located for purpose of calculating reaction force. Neither the mass of the earth nor the mass of the universe is required to be proactive in creating gravitational force... nor are individual masses sending out signals by gravitons, or any other form of traveling hypothesized physical transmission wrt space. It is (resistence to disassociation) of the earth's inertial mass, that creates the initial pressure deficit (proportional to earths mass). The increased negative pressure creates the gradient by which other bodies feel the negative pressure presence of the earth....and by which a collective cosmological pseudo force emerges in form as the added inertial reaction of all other matter in gravitational communication with the earth through the auspices of expanding space. Earth's mass and cosmic mass delineate only as reactions to the isotropic acceleration field created by spatial expansion. Spatial expansion powers the universe.

---

[12]Which raises the question as to whether 'g' fields eventually terminate. As electrical fields are presumed to ultimately find an opposite charge, can the same be presumed for gravity? Peculiar behavior of pendulums have been reported during lunar eclipse.

The cosmos thus has two operative modes.[13] Isotopic reactionary fields (gravity) are created by isotropic spatial expansion, whereas parallel constant intensity fields arise from the reaction of the universe to directionally accelerated masses. The notion of reactionary inertial space can be analogized to a piston of mass M_p inside an imaginary cylinder punched through the universe (shown dotted blue in Fig **XVI** and **Fig XVIII**). The spatial density ρ_U is taken as (3×10^{-26} **kg/m^3**), and consequently the effective area density from (P-2) is $\rho_U R/3$. The piston mass M_P has two modes of communication with all other inertial bodies scattered throughout the cosmos. From a gravitational perspective, M_P sees the universe as a spherical spatial surround lightly sprinkled with chunks of matter. As an accelerating inertial mass M_P, the universe exists as a plenum of planes which collectively contain the entire mass of the universe as an effective area-density confronting the acceleration of M_P. As previously stressed, reactionary forces are always opposite to the direction of acceleration, consequently the pseudo force acting upon each element, atom or subatomic particle, will be anti-parallel to the direction acceleration. Each such element will have the reactionary impedance of the universe imposed thereon as shown in **Fig XI**.[14]

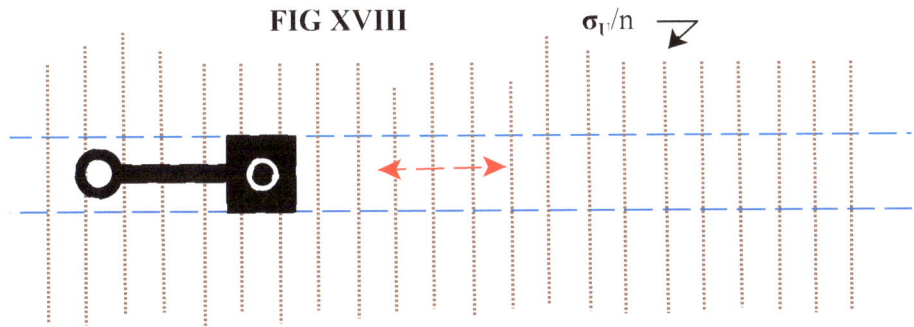

While at rest, continuous communication between the piston and all other mass in the universe is maintained in accordance with the 3-D isotropic expansion mode (green) depicted in **Fig XVI**. When the piston accelerates in the cylinder (red arrow as shown in **Fig XVIII**), that part of the cosmos that acts upon the piston can be represented as as a plenum of flat area densities (dotted brown) each having an area density σ_U/n The effective 'g' field acceleration parallel to the axis of the cylinder, and contained therein, is thus $\sigma_U \times a_n$. Within the cylinder, the **g** field of all such area-density planes manifest as cumulative, i.e., the gravitational field of each area density plane is parallel to all other area density planes and therefore pressure is additive. Each fractional area of every plane contained within the cylinder contributes equally to the total pressure irrespective of whether the plane actually incorporates a physical mass within the area of the cylinder.

---

[13] That which Einstein long ago perceived as the equivalence principle, removes all dissimilarities between inertial and gravitational mass. What was left unrevealed, is that their remains for explanation, the operative difference between the way the universe functions in the two applications.

[14] This even though by chance, the punched out cookie cutter cylinder may have no physical matter-energy within its entire length -- i.e., the cylinder can be considered empty, the contents thereof having been removed by the cookie cutter punch-out.

The area density planes are virtual - they do not connote the presence of real mass anywhere. As perceived by the piston, however, the local area density planes passing through the piston appear to be the source of a '**g**' field that is much larger than that can be explained by the local area density σ_U/n of an individual plane. But the parallelism of the field gives no indication as to its distance of the source - from the perspective of the piston, there is a mass deficit.

The cumulative '**g**' field that results from the parallelism of directional acceleration, is the result of the on-going expansion of the negative pressure universe - Force fields are pressure fields, but expansion of a negative pressure volume is not an easy proposition. The bulk modulus of space will be infinite or nearly infinite - consequently the velocity of propagation for a longitudinal wave in a negative pressure environment will be infinite. Accelerations '**a**' instantly couple to the universe as evidenced by the immediacy of inertial reactions. Moreover, they are simultaneously instantly opposed by every laminate as shown in **Fig XVIII**. It can now be understood why inertial reactance pressure ubiquitously exhibits as a counter pressure $(\sigma_U)\mathbf{a}$. Every directional acceleration is instantly opposed by every laminate into which the universe can be sliced. The pseudo force '**g**' fields associated with the inertial affects of masses upon the cosmological expansion field, are likewise instantaneously dependent upon inertial reactions, and therefore immediately extinguished when a mass is converted to another form of energy.

Recapitulation

Local Inertia + Spatial Expansion —> Gravity

Within what is known of the extent of space and the mass of matter, the cosmos is a superlative vacuum. Yet an accelerated or decelerated mass in free space feels cosmic presence in some fashion as opposition. By what manner can space bring about gravity, inertial reaction and the anomalies of propagation?

Surprisingly, the existence of such an artifice has long been known, but commonly viewed as having little practical applicability – indeed, the miraculous properties of infinite laminae(s) were considered largely academic. That began to change in the latter years of the 20th century. Coming with the 1998 supernova studies, were the imperatives of exponential expansion, infinite space and the problem of finding sufficient energy to fund accelerated expansion.

While neither Hubble mass nor size is separately utile, taken together they define the scalar density semblance of the universe. To appreciate the simplicity of gravity, it is first necessary to expose the subtlety of inertia, for that is the stuff from whence it emerges.

FIG 1

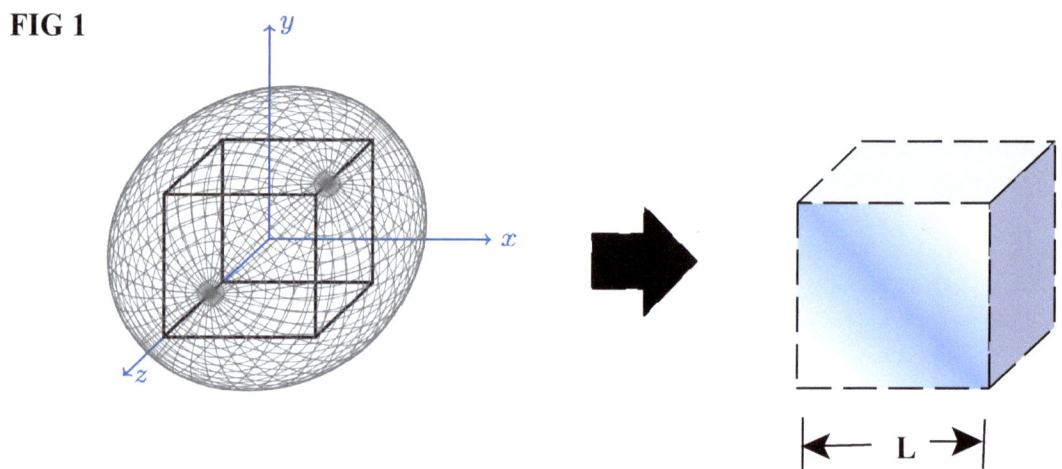

Fig 1 represents the Hubble sphere. The average volumetric density ρ_U is in the range of:

$$\rho_U \approx 3 \times 10^{-26} \text{ kg/meter}^3 \qquad (1)$$

The Hubble radius is on the order 1.3×10^{26} meters (corresponding to a Hubble constant 70). Taking the Hubble as a sample of a universe that is at least as large as the Hubble sphere a (perhaps even infinite), we then pluck out a cubical volume of mass **M** having a side length $L = 10^{26}$ meters. The object will be to find an equivalent area density σ_U in each coordinate direction that corresponds to the volume density ρ_U. The cube has six sides and each has an area L^2. Since the volume density ρ_U equals M/L^3 then for two parallel planes orthogonal to each coordinate axis:

$$M = \rho_U[L^3] = 2\sigma_U[6L^2]$$

And therefore:

$$\sigma_U = [L/3][\rho_U] \approx 1 \text{ kg/meter}^2 \qquad (2)$$

Fig 3 shows opposite faces merged into three intersecting orthogonal planes. An accelerating body at the **x,y,z** origin sees the cube as an area density σ_U of infinite extent in every direction. Accelerations create equal and opposite counter forces confined to the area projection of the body upon the cosmic inertial plane per **Fig 4**.

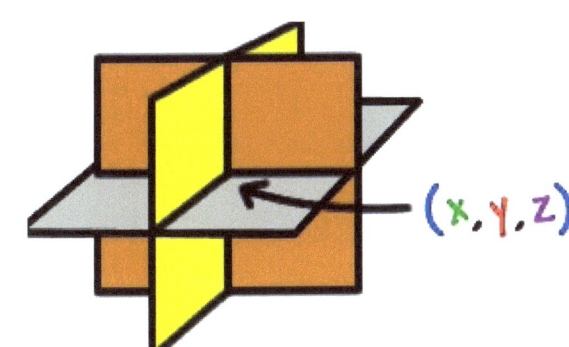

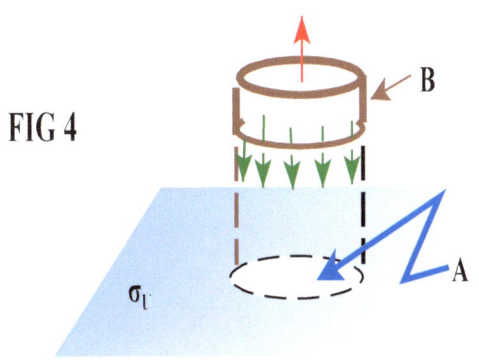

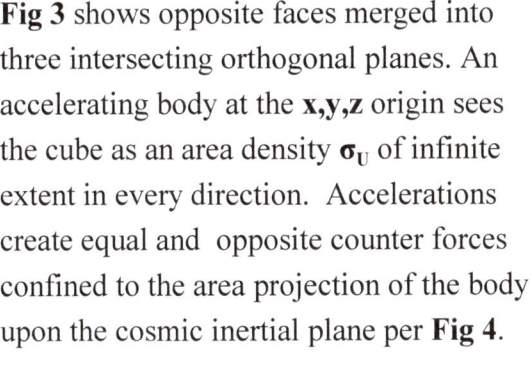

Fig 4: A cylindrical Mass **B** accelerated upward (red arrow) at rate a_1 creates parallel reactionary forces (green) upon all items of mass-energy contained therein. Momentum flow $[P = -(a_1)\sigma_U]$ (expressed in units of negative pressure) is inward negative, into the σ_U plane that represents the area density of the cubical block extracted from the Hubble sphere. Area **A** is the projection of **B** on the plane σ_U. Reactionary force lines are parallel, ergo, the intensity of the reactive field (green) is constant and (independent of distance).

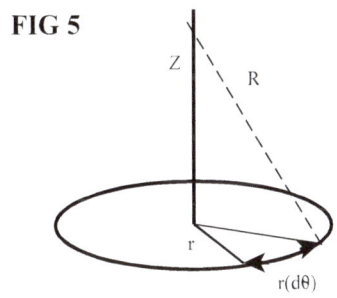

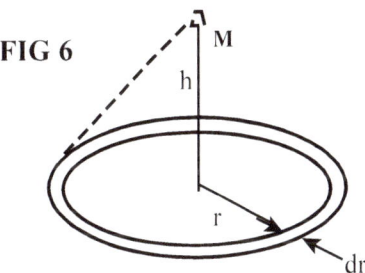

Fig 5 shows the geometry for finding the infinite flat plate potential at a point P above a uniform density disk having a radius '**r**' which is allowed to become infinite. The gravitation potential Φ is:

$$\Phi = -G\int_S \frac{\sigma(ds)}{R} = 2\sigma\pi G \int_0^a \frac{r(dr)}{(z^2 + r^2)^{1/2}}$$

Whence, when $r \gg z$, the potential Φ will no longer depend upon '**r**,' therefore:

$$\Phi = 2\sigma\pi G z \quad (3)$$

Force lines will be perpendicular to the disk, hence the imaginary Gaussian surface for calculating flux emanating from a gravitating body reduces to the area **2A** of the end faces of a cylinder axially perpendicular to the σ_U plane as shown in **Fig 7**. At any height '**h**' above σ_U, **Fig 6** applies to calculate the '**g**' field acting upon mass **M** by integrating the force created by all concentric rings from radius zero to infinity.

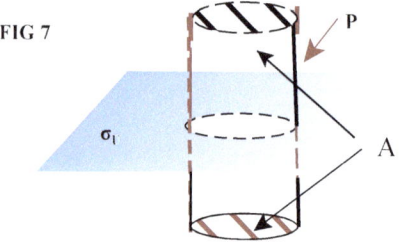

FIG 7

In the usual case where gravitational reaction is the result of the attraction of two approximately spherical bodies for one another, Newton's Gravity Law gives a good approximation, which can then be equated to Newton's 2nd law to calculate the gravitational acceleration:

$$\mathbf{F_g} = \frac{\mathbf{MG(M^*)}}{\mathbf{r}^2} = \mathbf{M^*(a)} \tag{4}$$

Since **M*** cancels on both sides of (4), all bodies fall at the same rate in a '**g**' field:, hence:

$$\mathbf{g} = \mathbf{a} = \frac{\mathbf{MG}}{\mathbf{r}} \tag{5}$$

However, for an infinite plane using the above development, the acceleration due to gravity is:[15]

$$\mathbf{g} = 2\pi \mathbf{G}\sigma_U \tag{6}$$

And therefore the gravitational force at any height '**h**' above the σ_U plane is:

$$\mathbf{F} = \mathbf{M^*a} = 2\pi \mathbf{G}\sigma_U[\mathbf{M^*}] \tag{7}$$

The 'g' force exerted by an infinite plane depends upon **M***. It is known that the '**g**' force exerted by a uniform density plane is constant. If all masses in the field of a plane fall at the same rate of acceleration, then the inertia of **M*** must also be independent of the height (distance between σ_U and **M***). That this be true, the inertial property of mass is also independent of distance from σ_U.

---

[15] For an infinite plane having an area-density **σ**, the **g** field at a perpendicular distance **r** is obtained from Gauss's law for gravity:

$$\int \mathbf{g} \cdot \mathbf{n} \, d\mathbf{A} = -4\pi \mathbf{GM}$$

since the '**g**' field will be perpendicular to the plane, the enclosing Gaussian surface can be a pillbox since only the area of the end areas **A** contribute to the flux integral. Then:

$$-\mathbf{g}(2\mathbf{A}) = -4\pi \mathbf{GM}$$

M is enclosed by the cookie cutter pillbox, so it has a mass **σA**, hence the **g** field is independent of distance:

$$\mathbf{g} = 2\pi \mathbf{G}\sigma$$

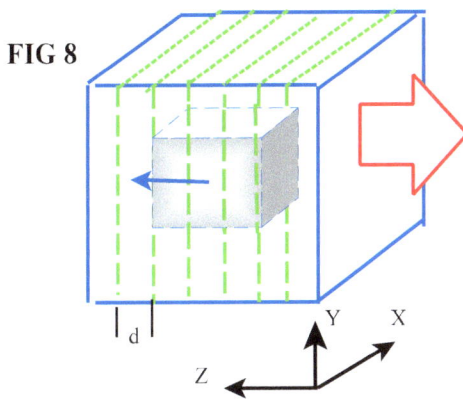

FIG 8

A body **B** in uniform motion wrt the universe, or vice versa, experiences space as a superlative vacuum (ρ_U). But changes in magnitude or direction, are precipitously moderated by counteraction. Meditating as area-density, the cosmos functions as a plenum of infinite planes which take effect cumulatively as a ubiquitous area-density σ_U.

Both Newton's 2nd law and his law of Gravity are founded upon the *(mass) x (acceleration)* product. Understanding inertia is precedent to understanding gravity.

$$\mathbf{F_I = Ma}, \qquad \mathbf{g} = \mathbf{M}\frac{G}{r^2} \qquad (8)$$

Deeming the first expression applicable to the cosmos as a whole, both sides are divided by meters². With $(\mathbf{a}^2)(\mathbf{M_u})$ substituted for $\mathbf{F_I}$, the cosmic inertial counter pressure reduces to:

$$\mathbf{a_2}\frac{\mathbf{M_u}}{\mathbf{m}^2} = \frac{\mathbf{M_B}}{\mathbf{m}^2}\mathbf{a_1} \qquad (9)$$

Equation (9) restates what has previously been developed, that is, momentum flow into an accelerating mass must be counter balanced by cosmic inertial counter pressure (momentum flow from the universe). Formulating the cosmos as an operative area-density impedance, greatly aids the comprehensibility of inertia. Pressure is momentum flow - but spatial pressure acting upon mass cannot be analogized to kinetic action of gas molecules rebounding from the walls of a container. For gravitational fields, spatial pressure educed by expanding inertial space provides the impetus for the reactionary fields of local bodies. Hence, $\mathbf{g}$ can be substituted for $\mathbf{a_2}$ and the cosmological acceleration factor $[\mathbf{a_n}]$ can be substituted for $\mathbf{a_1}$, thence (9) becomes:

$$\mathbf{g} = \frac{\sigma_B}{\sigma_U}[\mathbf{a_n}] \qquad (10)$$

where σ_B is the area density of a uniform mass $\mathbf{M_B}$ and $\mathbf{a_n} = \mathbf{c}^2/\mathbf{R}$ is the acceleration factor applicable to an exponentially expanding cosmos in a $\mathbf{q} = -\mathbf{1}$ universe.[16]

The local **g** field of a mass **M** is the result of four concurrences:

1) Expansion created isotropic spatial acceleration.
2) The pseudo field created by inertial opposition of **M** opposing (1).
3) The action of the pseudo field (2) upon the universe.
4) The reactionary field of the universe created by (3) acting upon **M**

---

[16] The deceleration parameter (**q**) expresses the change in the radial rate of expansion in terms of Hubble parameters. For an accelerating universe, **q** = -1, and the rate of acceleration is $\mathbf{c}^2/\mathbf{R}$.

Spatial growth as the ideate of positive energy created by the expansion of negative energy, reveals the essence of self creating cosmology. Expansion of a negative pressure volume creates positive energy.[17] That this condition is also an essential element of the zero energy universe, suggests a formulation. Specifically, if the distance between opposite faces (**Fig 1**) is increasing at an accelerating rate c^2/R, the negative pressure **-P** is:

$$-P = (\sigma_U)a_n = (c^2/R)(\sigma_U) = (c^2/R)[M_U/4\pi\sigma_U R^2] \qquad (11)$$

However, from 2nd Friedmann's equation (with Λ), the universe will have zero energy if:[18]

$$-P = \rho_U c^2/3. \qquad (12)$$

Substituting $(M_u)/(4/3)\pi R^3$ for ρ_U, (12) is seen to be identical with (11). Consequently, Friedmann's equation reduces to:

$$\ddot{R} = \frac{\Lambda R}{3} \qquad (13)$$

Einstein introduced the cosmological constant Λ to balance gravity. This required Λ equal $3H^2$, in which case (13) has the solution: de Sitter's empty universe and the zero energy universe (Positive Mc^2 energy = negative gravitational energy) have the same solution:

$$R = R_o[e^{Ht}] \qquad (43)$$

Ergo, a zero energy universe expands exponentially. And per (40), negative '**g**' pressure created by exponential expansion ($a_n = c^2/R$) will be equal to the positive **M**.

---

[17]First proposed by William McCrea (circa 1959), an expanding negative pressure creates positive energy. In the context of the present development, this translates to inertial accretion rather than new particles by way of the now discredited "*steady state theory.*" Initially, the idea of new matter was seized upon Fred Hoyle and other advocates of steady state theory, to explain the expanding universe as a constant density proposition. The concept is also the basis for the inflationary theory first put forth by Alan Guth in ( ). Both Steady State and Inflation were founded upon $P = -\rho_U c^2$ as the determinative equation of state. But a theory based upon the presumption that density remains constant during expansion, is bootstrap. There is no other reason to assume density is constant other than for the sake of the theory founded thereon. The zero energy universe by contrast, requires negative and positive energy always balance to zero. The equation of state is $-3P = \rho_U c^2$. In lieu of rapid doubling, the universe experiences an early stress phase when the expansion rate $c^2/R \longrightarrow \infty$ for an infinitesimal duration when $R \longrightarrow 0$.

[18]The 2nd Friedmann equation with the cosmological constant is

$$R = \frac{4\pi G}{3}\left[3P + \rho_U c^2\right]R + \frac{\Lambda R}{3} \qquad (44)$$

While our theory of inertia and gravity does not improve upon the accuracy of the empirically established relationship between space, mass and acceleration, neither do the experimental findings conflict therewith. While Hubble density has yet to be precisely established, both the cubical and spherical constructs of inertial space are consistent with the amalgamation of estimated Hubble mass diluted over Hubble area as one **kg/m²**. That the ratio of Hubble mass to Hubble area (in the mks system of units) is approximately unity within the limits of experimental error, the question arises as to whether σ_U can be shown ≡ "**one**." If so, the collective influence of all other mass must be consistent with the known symmetries of the universe as a whole.[19]

Admittedly, ubiquitous virtual inertial area-density is offensively counterintuitive. Yet for masses uniformly spread over large area planes, the gravitational consequence follows straightaway. Should not the parallelism of field lines for directional accelerations lead to the same result for inertial reaction if space is non-compressible. Should the pseudo force created by the acceleration of mass wrt space be any different than the pseudo force created by an accelerated space?

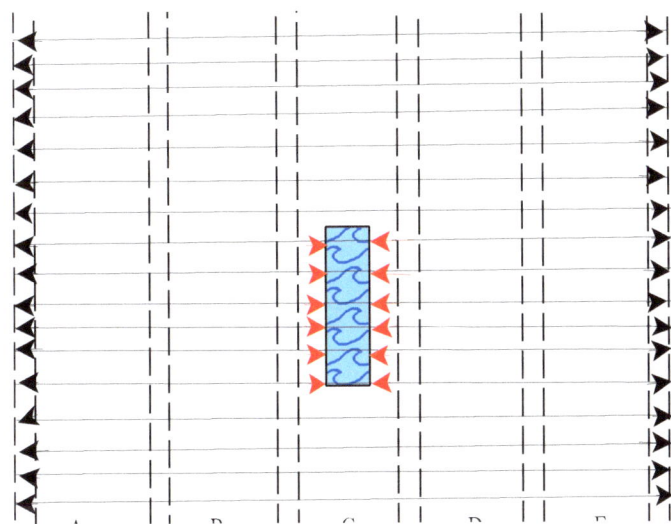

Fig 9: Spatial expansion modeled as 2-D bidirectional acceleration divergence (black arrows) with area-density slabs separated by dotted lines. Earth depicted as transformed to a slab of area density σ_B (blue) with inwardly directed arrows (red) indicating pressure caused by the impediment of earths interior mass to internal spatial expansion. Momentum influx pressure (earths '**g**' field) at the earth's surface is defined by the projection of earth's surface upon the area density σ_U of the universe

$$P = a_N \sigma_B \qquad (44)$$

Thus, if gravity were unidirectional, transformation of earth's mass to its surface then creates a shell density $\sigma_E = M_E/A_E$, Hence (44) would be a correct expression for the surface pressure field. However, isotopic expansion acts upon the individual elements of mass within the volume of a body, so the similitude as well as the distinction, between surface pressure **P** and '**g**' becomes significant. Per Einstein, it should make no difference whether mass is accelerated wrt the universe or vice versa. Taking M_E as **5.98 x 10²⁴ kg** and r_E as **6.37 x 10⁶ meters**, surface pressure **9.8 ntn/m²**. :

$$P = \frac{F}{A} = \frac{M_E(a_n)}{A_E} = \frac{\{5.98 \times 10^{24}\}(c^2/R)}{4\pi[6.37 \times 10^6]^2} \approx 9.8 \text{ ntn/m}^2 \qquad (45)$$

---

[19] Because of non-linear gravity acting upon gravity fields, transposing between 3-sphere and 2-sphere geometries, requires compensatory adjustments. So while Hubble bare mass M_b will be taken as 1.5×10^{53} kg. That the default value of the area density $\sigma_U = 1$ then requires a companion radial scale size R = 1.1×10^{26} meters. From (9), the σ_U value of the universe follows from Newton's 2nd law - the calibrated using the earth as a test mass.

Calculating force on a make-believe area-density, outputs the answer in units of pressure. While a desirable form for explaining momentum flow in the formulation of gravity as well as pseudo forces arising from directional accelerations, in the real world, gravity is not surface pressure, but rather momentum flow across the space-mass interface that dissipates throughout the volume. This is the cosmological response that counteracts the expansion deficit create by internal mass.

Outputting '**g**' fields (and other pseudo forces) in standard form (**ntn/kg**), requires a perceptual take on the cosmic response to the affect mass on the expansion field. In magnitude and dimensionality, the response to acceleration, is a counter acceleration that preserves the uniquity of each mass in relation to all other mass-energy throughout the universe. The reaction of the earth's inertial mass to the primary field (isotropic spatial expansion) creates a negative pressure sink which affects diverging space (sensed by the expanding area density shells) made manifest dimensionally as deceleration $[g(\sigma_U)]$. That local inertia can be successfully related to global inertia, the (σ_U) factor must = **1 kg/m²**. In essence, a restatement of *Mach's Principle*. Any other causal determinative would violate Newton's 2nd second law. Suffice to note, cosmic counter action marshals the inertial content of the universe in the form of a longitudinal pressure wave $(\sigma_U)g$.

Fig 10 illustrates the operative cosmos for 3-D expanding space. Inertial slabs have been replaced by inertial shells, but the concept of inertial space as a scalar area density σ_U is retained. Field intensity of earth's spatial pressure field '**g** σ_U' falls off inverse squared. The inertial reaction at each shell is retro-directive, convergent upon earth. At the Hubble limit **R**, the reflective density along any line of action = σ_U.

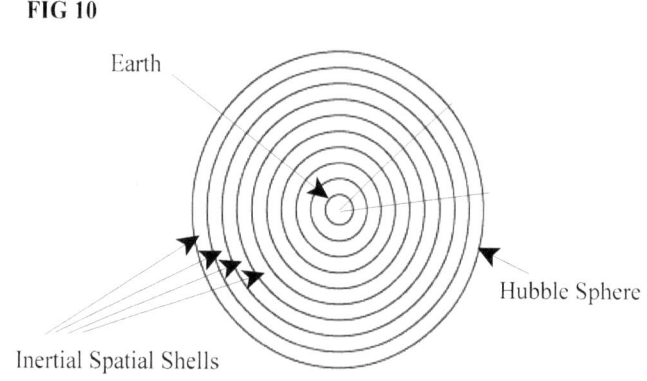

FIG 10

Earth

Hubble Sphere

Inertial Spatial Shells

The projection of a pie shaped section of earth extended to the Hubble limit, illustrates how the intensity of the expansion pressure dilutes with distance. All cosmic mass can thus be considered distributed over the Hubble surface at distance **R**. As is the case for unidirectional accelerations, isotropic expansion creates equal and opposite forces between the universe and its constituent parts.

Taking the earth as an approximately uniform spherical mass density σ_E, the surface pressure due to expanding inertial space is:

$$P_E = (a_n)(\sigma_E) \tag{46}$$

and the pressure on the universe is:

$$P_U = (a_2)(\sigma_U) \tag{47}$$

For net momentum flow = zero, the two pressures will be equal, therefore:

$$\mathbf{a}_2 = g = \mathbf{a}_n \frac{\sigma_E}{\sigma_U} \tag{48}$$

From Newton's Law of gravity, the intensity of the 'g' field at a distance 'r' from center of a spherical mass M_B is:

$$g = M_B G/r^2 \tag{49}$$

Thence, combining (48) and (49):

$$G = \frac{r^2 g}{M_B} = \frac{r^2 \sigma_B [a_n]}{M_B \sigma_U} = \frac{r^2 M_B [a_n]}{4\pi r^2 \sigma_U M_B} = \frac{[a_n]}{4\pi \sigma_U} \tag{50}$$

We are now in a position to calculate G from Hubble parameters. Two factors need to be determined. The cosmological acceleration factor 'a_n' can be expressed in terms of the Hubble scale R and the rate of recession of the Hubble surface. Specifically, radial rate of change is obtained from the "so called" deceleration parameter 'q' defined as:[20]

$$q = (-)\frac{\ddot{R}(R)}{(\dot{R})^2} \tag{51}$$

Since it is now widely accepted that the expansion rate is increasing, 'q' will have a negative sign and value of unity, i.e., $q = (-1)$, accordingly:[21]

$$a_n = \ddot{R} = \frac{\dot{R}^2}{R} = \frac{c^2}{R} \tag{52}$$

R is the operative value of the spatial area-density arrived at by transformation of Hubble bare mass from volumetric density to a surface integral (3-sphere —> 2-sphere). Hence (50) can be written as:

$$G = \frac{c^2}{4\pi R \sigma_U} = \frac{c^2 R}{M_U} \tag{53}$$

That the first expression for G encodes the cosmological acceleration factor c^2/R, follows from its dynamic dimensionality $(m^3/sec^2)/kg$. That it is inversely dependent upon the operative scale 'R' imposes a reciprocality upon the inertial aliquot of M, in that the MG product be constant in an expanding universe. In the 2nd expression, M_U is substituted for the Hubble mass $4\pi R^2 \sigma_U$, per (11) the condition that Hubble that the ratio of Hubble area to bare mass equals '1.' M_U increases as R_2.

---

[20]Misnamed "*deceleration parameter*" at a time when cosmologists were convinced expansion would be slowed by gravitational effects. For an accelerating cosmos, $q = (-1)$

[21]It should not be surprising that G, having dimensionality meters^3/sec^2, encodes the cosmological acceleration factor c^2/R. That the format of (P-8) comports with the acceleration created by a rotational velocity 'c' at radius 'R' will prove to be of value in understanding the nature of expansion.

Provinciality of Inertial Reaction

As evidenced by the inertial reactance imposed upon accelerating masses, pseudo forces are retro-directive, independent of the location, orientation or composition thereof. Even at subatomic dimensions, there is no identifiable line of action between an accelerated particle and its connection to the universe. No part or portion of any entity participating in the reactance of the whole can be distinguished incrementally from any other part - each elemental volume of an accelerating body, blends seamlessly into the reactionary force of the whole with no apparent distinction for the quantified nature of subatomic particle masses That the reactance of an accelerated body can be expressed in terms of an infinite number of cross-sectional planes orthogonal to the direction of acceleration, leads to the following lemma:

> A unidirectionally accelerating body modeled as a plurality of N planes orthogonal to the direction of acceleration, will experience the same total reaction irrespective of the location or orientation of any of the planes.

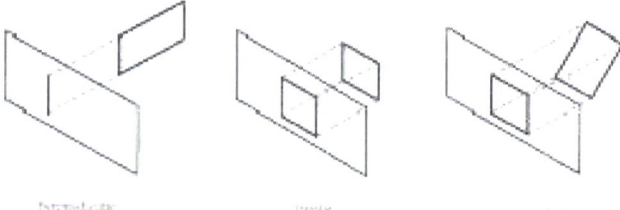

Fig 11. A parallel plane (center) sliced from an accelerating body experience the same inertial reactance irrespective of its later orientation (edge-on or inclined) with respect to the σ_U plane defined by the direction of acceleration. .

Taking the cube universe as a single infinite plane perpendicular to each coordinate axis, effectuates the imposition of σ_U upon each of the **N** planes into which the accelerating body is divided. In the limit (as $N \longrightarrow \infty$), all planes can be considered coincident with the virtual area density function σ_U as shown in **Fig 12**. There are no lines of force connecting the inertial property of accelerating bodies to the universe, nor any element thereof. The universe exists ubiquitously as virtual area-density σ_U coincident with each element of mass.

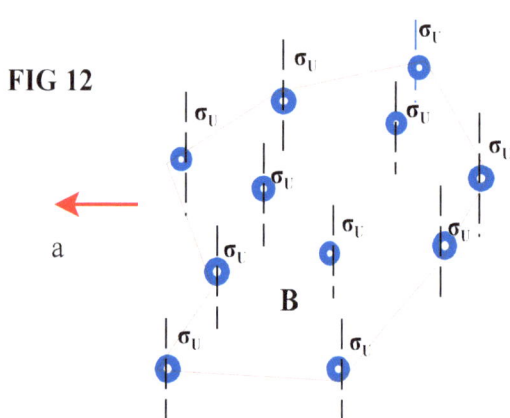

FIG 12

Fig 12: An accelerated body **B** comprising 11 particles held together by electrical and quantum forces. While all non-accelerating masses experience the universe as a near perfect vacuum, accelerating entities perceive the universe as virtual area-density σ_U. Each particle of '**B**' is instantly a pawn in the cosmic game played by the universe in the guise of a superimposed area-density σ_U. As the local representative of cosmic mass, the σ_U inertial field [kg/m^2] renders unto directional accelerations that due according to their mass.

Space As An Incompressible Ideal Fluid

Studies of space as a propagation medium and spacetime as an intermediary for the propagation of gravitational waves, invariable spend considerable effort in showing by mathematical means, (rich with assumptions and presumptions), the nature of space to be that of a perfect fluid. For didactical purposes, it will be useful to make a case for space as the quintessential medium.

Fig 13 illustrates the set-up for a piston accelerating to the right in a cylinder formed by boring a hole through the universe. If the piston moves at constant velocity "**v**" it encounters no resistance from the spatial medium. However, the slightest change in the piston's velocity wrt the universe, creates immediate inertial consequences. The question posed, is how fast does the inertial reaction pressure move ahead of the piston, for that determines how much of the universe participates in creating the inertial force. The bulk modulus β for real fluids is

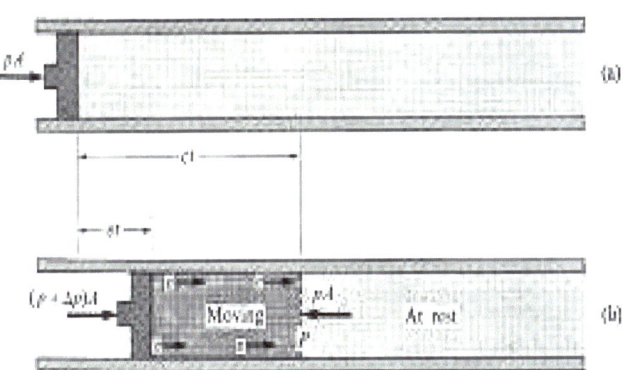

$$\beta = \frac{\text{change in pressure}}{\text{fractional change in volume}} = \frac{\Delta P}{\Delta V/V} \tag{54}$$

From which longitudinal velocity 'v_L' is:

$$v_L = \sqrt{\frac{\beta}{\rho_U}} \tag{55}$$

The volume of an incompressible fluid does not change with pressure. There being no change in volume, spatial bulk modulus $\beta = \infty$, and consequently, so also does 'v_L.' Changes in the velocity status of a mass, or any '**g**' fields derived therefrom, can be expected to instantly communicate with distant masses as longitudinally transmitted pressure displacements. That cosmic response to the motion of the piston is instantaneous, it must of necessity, represent the participation of all planes comprising the composite area-density σ_U as shown in **Fig 14**. Resistance to acceleration is an instantaneous property of inertial mass, born in the virtuality of localized cosmological counter action. Though an idealized aspect of space as an incompressible fluid, pressure in form of momentum flow, is nonetheless real. By contrast, gravity waves, like EM waves, are known to be transverse. The propagation velocity of transverse

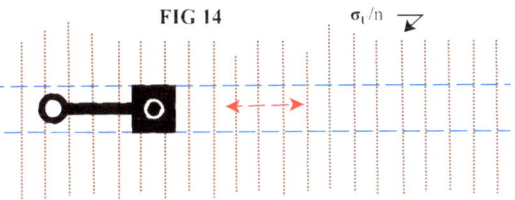

waves depends only upon the square root of three times the pressure divided by density. Pcr (44), the pressure (-P) in the zero energy universe, must equal $\rho_U c^2/3$ Therefore:

$$v_T = \sqrt{\frac{-3P}{\rho_U}} = \sqrt{\frac{\rho_u c^2}{\rho_u}} = c \tag{56}$$

Transverse wave velocity depends from cosmological pressure and density. They propagate at the velocity of light, 'c.' As is known, the same result obtains from electrical properties of free space:

$$c = \frac{1}{\sqrt{\mu_0 \varepsilon_0}} \qquad (57)$$

Which taken together with the electrical impedance **Z** of free space and the inertial area density, determined from Hubble parameters, relates the electrical and mechanical formalisms in terms of a conversion constant **K**:

$$Z = \sqrt{\frac{\mu_0}{\varepsilon_0}} = K\sigma_U \qquad (58)$$

$$c^2 = \frac{-3P}{\rho_U} = \frac{1}{\mu_0 \varepsilon_0} \qquad (59)$$

$$3\sigma_U = R[\rho_U] \qquad (60)$$

Accelerations communicate with, and are referenced to, the universe as a unified area-density whole. For every directional acceleration, the cosmic impedance σ_U is locally present. While area density can be modeled as a laminate coextensive with the universe, an accelerated body feels the yoke of opposition of all planes combined. A cubic sample of average density selected anywhere within the Hubble can be shrunk to a vanishing small "area-over-volume-ratio" applicable to the divergence of the cube as a whole. Within the area defined by the cross sectional dimensions of an accelerating body, the distended field in both the forward and rearward direction measured with respect to the acceleration, are tantamount to a bidirectional gravitation field, whence the same logics apply to the cumulative effect at any point. Taking a cubic area of volume 'V' and shrinking opposite sides equally to reduced the volume to line while maintaining the same density ρ_U, then:

$$\mathbf{div} = \lim_{V \to 0} \frac{\iint_S \mathbf{a} \cdot \mathbf{n} \, dS}{V} = \frac{\mathbf{a}[L^2]\rho_U}{RL^2} \times \frac{1}{3} = \mathbf{a}\sigma_U \qquad (61)$$

That (61) defines the area-density impedance pressure $\mathbf{a}\sigma_U$ at any point along a line of action coextensive with the cosmic dimensionality **L = R**, the spatial fluid must either be incompressible or otherwise effectual in transmitting a longitudinal pressure wave at superluminal speed.

In Conclusion

Both Gravitational and EM waves are quadrupole transverse. And while the quadrupole angle for gravity waves is $\pi/4$ and the electric angle is $\pi/2$, the forms of transmission are otherwise governed by the physics applicable to transverse finite wave propagation per (56). For EM waves, speed of propagation depends (per James Clerk Maxwell), upon the permeability μ_o and permittivity ε_o of the vacuum. For gravity waves, propagation speed depends upon the spatial pressure (-3P) and density ($\rho_U c^2$) factors, befitting a zero energy universe *a la* Friedmann's 2nd equation (44).

Inertial reactions and their consequent '**g**' fields, however, are not the result of transverse wave prorogations. They are, rather, manifestation(s) of isotropic spatial expansion. Expansion of negative pressure volume creates positive energy at the expense of increased spatial rigidity. A universe in tension hardly resists further expansion. The un-expansible state of negative pressure space corresponds to no volume change with pressure. That cosmic pressure must by negative for a zero energy universe per Friedmann's second equation (44), the spatial bulk modulus β of negative pressure space can be written as:

$$\beta = \frac{\text{change in negative pressure}}{\text{fractional change in volume}} = \frac{\pm \Delta P}{\Delta V/V} \tag{62}$$

For a longitudinal pressure wave, the propagation velocity 'v_L' is:

$$v_L = \sqrt{\frac{\beta}{\rho_U}} \tag{63}$$

Pressure is negative for a universe in tension. Consequently $\Delta V \approx$ **zero** irrespective of the direction of change ΔP. Hence $\beta \longrightarrow \infty$. Likewise, so also '$v_L$' $\longrightarrow \infty$. All forms of acceleration are immediately communicated to the universe and reactionary counter pressures are summoned for immediate application.

The state of space as incompressible negative pressure renders the universe intelligible. While transverse waves propagate at the speed of light, force fields are not the ephemeral creations of a central source that require perpetual updating. No messenger particles are required and no maintenance updating is needed to explain natures long range force fields. Communication is in form, momentum flow between the collective area density σ_U and the individual masses that make up σ_U. Local '**g**' fields are not internally created enigmas initiated within masses, but rather the observed manifestation of momentum influx. Inertial opposition of masses to the acceleration field engendered by expanding space, creates pressure deficit sinks. That the pressure deficit imbalance $a_n \sigma_B$ is precisely corrected by the compensatory inflow (**g**)σ_U, completes the description of gravity.

www.ingramcontent.com/pod-product-compliance
Lightning Source LLC
Chambersburg PA
CBHW040433220526
45473CB00004B/1428